INVASIVE SPECIES: DETECTION, IMPACT AND CONTROL

Invasive Species: Detection, Impact and Control

Charles P. Wilcox
and
Randall B. Turpin
Editors

Nova Science Publishers, Inc.
New York

For permission to use material from this book please contact us:
Telephone 631-231-7269; Fax 631-231-8175
Web Site: http://www.novapublishers.com

LIBRARY OF CONGRESS CATALOGING-IN-PUBLICATION DATA

Invasive species : detection, impact, and control / Charles P. Wilcox and Randall B. Turpin, editors.

p. cm.

ISBN 978-1-60692-252-1 (hardcover)

1. Nonindigenous pests. 2. Biological invasions. I. Wilcox, Charles P. II. Turpin, Randall B.

SB990.I575 2009

628.9'6--dc22

2008038236

Published by Nova Science Publishers, Inc. ✝ New York

CONTENTS

Preface

Invasive species is a phrase with several definitions. The first definition expresses the phrase in terms of non-indigenous species (e.g., plants or animals) that adversely affect the habitats they invade economically, environmentally or ecologically. It has been used in this sense by government organizations as well as conservation groups such as the IUCN. The second definition broadens the boundaries to include both native and non-native species that heavily colonize a particular habitat. The third definition is an expansion of the first and defines an invasive species as a widespread non-indigenous species. This last definition is arguably too broad as not all non-indigenous species necessarily have an adverse effect on their adopted environment. An example of this broader use would include the claim that the common goldfish (*Carassius auratus*) is invasive. Although it is common outside its range globally, it almost never appears in harmful densities. This new book presents important recent research in the field from around the world.

Short Communication 1 - Ailanthus (*Ailanthus altissima*) is an exotic invasive tree species that is becoming an increasingly common threat to native forest communities in the eastern United States. The species spreads aggressively by both seed and sprouts and produces allelopathic chemicals that may enable its spread. The authors tested two chemical control methods on Ailanthus (cut stump application with an aqueous 50% formulation of triclopyr vs. basal bark application using a 20% ester formulation of triclopyr). Each application was applied to replicated groups of trees in the spring following budbreak or in the fall before leaf fall. The cut stump application resulted in 100% control of *Ailanthus* (no resprouting) in both spring and fall, while the basal bark application was >90% effective (complete crown mortality). Spring applications were slightly more effective than fall applications for basal bark treatments. While chemical control appears to be an effective method for killing *Ailanthus* trees, abundant germination from the soil seed bank surrounding controlled trees will require long-term control efforts.

Short Communication 2 - Invasion of the Mediterranean Sea by alien species is getting increasingly prominent every day. These alien species are probably transported in with ballast waters or by attaching themselves onto vessels. Furthermore, many species have been brought into the Mediterranean for aquaculture purposes or inadvertantly introduced from public aquariums. But, most of the alien taxa recorded in the Mediterranean are originated from Indo-Pacific, indicating that the Suez Canal is considered to be the major vector in introducing alien species into the Levantine Basin. More than five hundered alien species have been recorded so far and each year new arrivals are reported (Galil, 2008). Alien invasion is considered to be a major problem worldwide, especially if the native coastal ecosystems are polluted and disturbed by the anthropogenic activities, such as observed in the Mediterranean. The aliens can deplete food sources, change the habitat structure and environmental conditions, in which the native species cannot survive. However, the aliens with an economical value constitute an important portion of the fisheries in the Eastern Mediterranean.

Short Communication 3 - Invasive alien species can have a detrimental economic impact on human enterprises such as agriculture, grazing, forestry and tourist activities. Invasive species have been identified as one the major threats to ecosystems and biodiversity, as well as human well-being. The main objective of our study is to determine whether regeneration of the exotic *Pinus pinea* is able to compete with the regeneration of the native *P. canariensis*. The study area is located in the Natural Park of Tamadaba, 1400 m asl., in the NW of Gran Canaria island (Canary Islands). Stems and regeneration of *P. canariensis* and *P. pinea* were mapped in five randomly selected plots where both species were planted together around 45 years ago. Densities and basal areas of both species were also recorded. A monitoring of the survivorship of seedlings of both species was carried out during two years. A group of individuals of *P. canariensis* were excluded from grazing to determine the effect of grazing in the survivorship of the species.

Although the dispersal ability of *P. canariensis* was more effective, once the individuals of *P. canariensis* and *P. pinea* had been established, there was not difference in survivorship. Also, we did not find differences in survivorship for individuals excluded vs. non-excluded from grazing.

Despite the stability of the exotic species, this can change with the introduction of a dispersal vector of the seeds, a squirrel, *Atlantoxerus getulus*, which was introduced in Fuerteventura (the closest island to Gran Canaria) with an estimated population of 1 million of individuals. Gran Canaria is also suitable for the establishment of this exotic disperser of *Pinus pinea*.

Applying a precautionary principle, control of the species will be recommended in order to avoid future problems of invasiveness of *P. pinea*, as has been found after the introduction of the dispersal vector with another squirrel in South Africa.

Chapter 1 - *Spartina alterniflora* Loisel., a native to the Atlantic and Gulf coasts of North America, had been deliberately introduced into many countries for the control of coastal erosion and land claim. But now, the species has extensively dispersed, and even broken out in some non-native habitats. Due to Allee effect, inbreeding depression, rapid adaptation and evolution occur in the process of invasion and natural dispersal of *Spartina alterniflora*, it has become a model plant for studying biological invasion from both ecological and genetic perspectives. The previous researches showed that powerful ability of hybridization and introgression has been a genetic basic, superior reproductive capability has been the sources and strong ability of anti-stress and adaptability has been an ecological and physiologic basic for *S. alterniflora* invasion and expansion, respectively. In most invasive habitats, the expansion of *S. alterniflora,* based on intentionally transplants, indicated the mode of point dispersal. The episodic and continuous dispersal pattern of seeds has been playing an important role for maintain, recruitment and outbreak of *S. alterniflora* population. Meanwhile, consecutive expansion of *S. alterniflora* populations was ensured by the trait of potently clonal growth. Therefore, prevention of seed production in all designated areas is required to help contain this species and prevent its further spread. At the same time, although it has been proven to be very difficult, expensive and even impossible to eradicate the species, the integrated strategies with exploring the economical value and other restraining growth of *S. alterniflora* methods should be adopt to manage and alleviate the negative impacts the biological invasion.

Chapter 2 - Ascidians are common members of benthic marine communities. Due to their strong competitive abilities and their simple trophic requirements ascidians are highly invasive. Because they only need a hard surface for attachment and abundant particulate food to flourish, ascidians are easily introduced to new locations and can readily persist once established. Invasive ascidians often have considerable impact on invaded habitats. Not only can they affect benthic communities, but they also cause major problems for humans by overgrowing aquaculture equipment and organisms and by heavily fouling ships and man-made structures. Because ascidians have traditionally been of little direct economic value, much less is known about their biology than for other marine taxa (e.g., crustaceans, bivalves and teleosts). However, this is starting to change. Due to the impact invasive ascidians have had throughout the world in the past 25-30 years, ascidians have become the focus of significant scientific attention. Over the last few years a great deal of work has been conducted to learn more about ascidian ecology, to assess the impact of invasive ascidians on invaded systems, to prevent the spread of ascidians and to control them once they have become established in new areas. This review synthesizes the latest research on invasive ascidians and highlights areas for further study.

Chapter 3 - Infestations of wildlands by invasive plants can reduce resource productivity, decrease biodiversity, displace native vegetation, and alter ecosystem processes and functions. The traditional reactive strategy of controlling established invasive plant infestations followed by restoration of the native plant community has proven to be largely ineffective at reducing the spread and negative impacts of invasive plants. This approach often fails in its attempt to restore native plant communities and is too costly to apply at the scale required to have substantial effects. While large amounts of resources are intensively spent on efforts to restore a few infested wildlands, invasive plants continue to spread via emerging populations and expanding established infestations. A proactive approach with the objective of preventing new infestations and limiting the expansion of existing infestations is a more effective and efficient strategy for managing invasive plants in wildlands because it precludes the need for restoration. However, relatively few resources are being directed towards preventing the spread of invasive species. Successful strategies to prevent infestations of invasive plants should focus on: 1) limiting the spatial dispersal of propagules (i.e., reducing propagule pressure), 2) maintaining or increasing the ability of wildland plant communities to resist invasion (i.e., biotic resistance), and 3) systematically searching for and eradicating new infestations. Propagule pressure and biotic resistance interact to determine wildland plant community invasibility. At low biotic resistance even a few propagules may result in successful invasion; however, as biotic resistance increases, greater propagule pressure is required for invasion. Thus, efforts aimed at decreasing invasive plant propagule pressure and increasing biotic resistance can greatly reduce new infestations. Systematically searching for and eradicating new infestations is also a critical element of a successful prevention strategy, because uncontrollable events may still lead to new infestations. Successful management of invasive plants in wildlands will require more efforts and resources directed at prevention. This task can be facilitated by more research developing and improving prevention strategies and demonstrating the effectiveness of proactive management.

Chapter 4 - Resource managers can benefit from improved methods for identifying invasive plant species. The utilization of hyperspectral remote sensing as a tool for species-level mapping has been increasing and techniques need to be explored for identifying species of interest. The overarching objective of this paper was to investigate three distinct processing methodologies (i.e., Derivatives, Continuum Removal, and Shape Filter) to explore their potential for delineating wetland invasive plant species within the spectral domain of typical airborne hyperspectral sensors. Field-level hyperspectral data (350-2500nm) were collected for twenty-two wetland plant species in a wetland located in the lower Muskegon River watershed in Michigan, USA. Generally, continuum removed spectra were more similar than raw reflectance for the invasive species of interest according to the Jeffries-Matusita distance measure. Second-derivative analysis showed that the wavelength locations of absorption and reflectance features were consistent for all species and emphasized the NIR region for separation. The shape-filter was useful as a method to identify invasive species and showed that useful wavelength regions can vary depending on the species of interest and approach utilized. Using the shape-filter, *Lythrum salicaria*, *Phragmites australis*, and *Typha latifolia* possessed maximum separation (distinguished from other species) at the red edge (700nm) and water absorption region (1350nm), the near-infrared down slope (1000 and 1100nm), and the visible/chlorophyll absorption region (500nm) and red edge (650nm), respectively.

Chapter 5 - The authors have examined impacts of alien invasive plants on soil chemical properties, primary productivity and nutrient cycling in the plant / soil system. Specifically, they tested if impacts follow a general pattern across sites and species or, alternatively, if they are entirely idiosyncratic. The study first focused on 36 sites in Belgium invaded by one of the 7 most invasive plant species in NW Europe (*Solidago gigantea, Fallopia japonica, Senecio inaequidens, Heracleum mantegazzianum, Impatiens glandulifera*, *Prunus serotina* and *Rosa rugosa)*. The authors compared invaded to adjacent uninvaded plots for selected parameters. Primary productivity and nutrient uptake were always higher in invaded stands compared to uninvaded plots. Magnitude and direction of impacts on soil chemical properties strongly varied depending on site. However, impacts followed a general pattern, being predictable from soil chemical properties prior to invasion. Thus, in sites with low soil nutrient contents, invasion tended to increase available nutrient pools in the topsoil while the opposite trend was observed in soils initially rich in nutrients. This suggests that exotic plant invasion could lead to the homogenization of soil nutrient concentrations across invaded landscapes.

Later on, the authors selected two species (*Solidago gigantea* and *Fallopia japonica*) to study in details the mechanisms of the impacts on soil properties.

In soil invaded by *S. gigantea*, soil phosphorus availability was increased. Higher turnover rates of phosphorus in belowground organs and mobilization of soil sparingly soluble P forms through rhizosphere acidification may be involved in the observed differences in soil P status between invaded and uninvaded plots.

In grassland invaded by *Fallopia japonica*, the carbon and nitrogen cycling were deeply modified. Due to its higher lignin/N ratio compared to resident vegetation, *Fallopia* litter decomposed much more slowly and immobilized a large amount of inorganic N, reducing the availability of this element in soil. On the other hand, the internal cycling of N in *Fallopia* was found exceptionally efficient. Indeed, about 80 % of the N present in aboveground biomass in summer is translocated to the rhizomes before leaves abscission. This process makes the plant relatively independent from soil N mineralization and possibly contributes to the high productivity and invasive success of the species. In addition, *F. japonica* also impacted soil fauna communities. The density of invertebrates under the canopy of *F. japonica* was reduced and the composition of the community shifted from a typical grassland community to typical forest groups. These changes may be explained by a reduction of food diversity, a change in soil microclimate and in organic matter quality.

Chapter 6 - Increased input of nitrate and ammonium to ecosystems is mainly responsible for eutrophication, which is related to biological invasion of aquatic exotic plant species. In order to determine the effects of different nitrogen forms on its growth and physiological responses in eutrophic water, *Eichhornia crassipes* plants were grown for 28 days in 5mmol/L nitrogen-contained nutrient solutions varying in NH_4^+:NO_3^- ratio (100:0, 75:25, 50:50, 25:75, 0:100) in laboratory. The results showed that the NH_4^+:NO_3^- ratio of nutrient solution dramatically affected the plant performance of *E. crassipes* including relative growth rate, number of generated ramets, nitrate concentration in root and leaf. Furthermore, nitrate reductase activity increased with reduction of NH_4^+: NO_3^- ratio in culture solution demonstrating preferences of N sources as nitrate in *E. crassipes*. However, ammonium concentration and glutamine synthetase activity in leaf did not significantly change, and those in root significantly increased when proportion of NH_4^+ in nutrient solution increased indicating that *E. crassipes* plants are able to resist ammonium toxicity by regulation of ammonium transport and assimilation. In a word, the results suggested that establishment and expansion of *E. crassipes* was not likely to be limited by nitrogen forms.

Chapter 7 - Warm-season perennial legumes, as a plant functional group, hold considerable promise for use as forage plants in the warm, humid southeastern U.S., where infertile soils and low-protein forage grasses are common. This plant group is large with tremendous ranges in growth forms and propagation methods. Despite the overall promise, only a few species have been, are currently, or even appear to be potentially useful forage plants. The few species of this group which are widely adapted across the Southeast are also considered, at least by some, to be invasive. Included are a vine (kudzu, *Pueraria montana* var. *lobata*), a tree (mimosa, *Albizia julibrissin*), and a herb (sericea lespedeza, *Lespedeza cuneata*). Rather casual observation can quickly provide evidence of the invasiveness of kudzu. Determination of the invasiveness of sericea lespedeza depends to some extent on the evidence assessed. With mimosa, the question may be about forage value rather than invasiveness. Most of the seemingly unstoppable spread of kudzu is from rapid vegetative extension, while seed provide the means for increase in range of the other two. Simple single-application treatments of various defoliation procedures or even the most effective herbicide have either had minimal or rather temporary effects on kudzu and sericea lespedeza. Long-term strategies, rather than single control treatments, are required for success. Knowledge of basic mechanisms of plant dispersal and ecosystem susceptibility to invasion by sericea lespedeza are needed to allow appropriate decisions about control and prevention of problem populations of this species. Developing technology holds promise for future use of mimosa as an intensively managed forage legume despite invasiveness. The greatest potential forage use of introduced, perennial, warm-season legumes in the southeastern U.S. in general may well be through the use of grazing livestock as part of carefully planned control strategies.

Chapter 8 - Making simple improvements on traditional fishing gear and methods can be applied to develop effective eradicating tools that can be integrated with other strategies used to combat invasive aquatic species. Swimming crabs are the target choice; they travel widely by swimming, attached to the hulls of ships or as larvae suspended in ballast water. They represent a valuable fisheries resource and are consumed as food. They are however also carriers of pathogens that may damage fisheries and aquaculture industries, which results in the need of effective eradication methods.

Traps make good eradicating gear because they are light, inexpensive, small, easily stacked onboard, keep the catch alive, and permit the release of non-target organisms undamaged. They are passive gear, so the target organism has to be lured into the traps with the promise of food, shelter, company or even a mating partner.

Improving luring methods, such as common baits or pheromone emitting decoys, is just as important as research on new trap design for increasing the catch. Determining the taste preferences of the target crab, and making bait combinations of fish plus sugarcane have demonstrated a synergistic effect that doubled the catch. Alternative bait made from fish mince inside teabags was as effective as ordinary bait, and allowed enrichment of the bait with attractants and the use of fish byproducts as raw material, which reduces cost and conserves valuable food resources. Pheromones offer attracting potential, have been found in swimming crabs and are emitted by the live decoys employed in their trap fishery. This fishing method eliminates the non-target catch by attracting only crab conspecifics. Specially designed traps containing live decoy crabs were more effective than baited traps, and eliminated unwanted organisms that feed on bait.

When designing eradicating gear, the quantity of the catch is more important than its size or quality; consequently, traps must be fitted with smaller meshed netting to additionally retain juvenile crabs. Observations on crab behavior around different trap designs showed that they search larger areas around traps with oval bases than those with rectangular ones, and this gives them more access to the trap entrances. Open funnel entrances allowed entry of most crabs contacting the traps and were superior to tight slit entrances because being open they permitted escape, thus reducing the non-target catch and the negative impact that lost gear causes on the aquatic resources by ghost fishing.

Chapter 9 - The state of Florida has among the two worst invasive species problems in the United States. Besides the sheer numbers of established exotic species in Florida, many present novel difficulties for management, or have other characteristics making effective management extremely challenging. Moreover, initiation of management action requires more than recognition by experts that a potentially harmful species has become established. It also requires the political will along with concomitant resources and appropriate personnel to develop effective methods and apply them. The authors illustrate various aspects of the situation in Florida with examples of invasive vertebrates, the problems they pose(d), and management approaches to the problems.

The problems described include long-established widespread and destructive species requiring intensive localized management (feral swine, feral cats); recently established species with potentially severe repercussions, but no broad operational removal programs yet in place (Nile monitor lizards, Burmese pythons), highly prolific mammals that could rapidly invade wide areas without containment/eradication (Gambian giant pouched rats, black-tailed jackrabbits); recently established, potentially destructive birds that might still be eradicated (purple swamp hens); species where sufficient public outcry resulted in control programs (black spiny-tailed iguanas); and rapidly expanding aggressive species for which no practical management actions are available (northern curlytail lizard). A species subset is used here to exemplify in more detail the array of invasive vertebrate species situations in Florida, including routes of introduction, impacts, surrounding politics, and management actions. These examples not only demonstrate the breadth of the terrestrial invasive vertebrate problems in the state, but they also show the diversity in resolve and response among the many species and the motivating factors.

In: Invasive Species: Detection, Impact and Control
Editors: C.P. Wilcox and R.B. Turpin
ISBN 978-1-60692-252-1

Short Communication 1

COMPARISON OF METHOD AND SEASON OF APPLICATION FOR CONTROL OF THE INVASIVE TREE SPECIES *AILANTHUS ALTISSIMA* IN VIRGINIA, USA

T.S. Fredericksen**, *M. Thurman, J.D. Fiore and D. Evans
Ferrum College, 212 Garber Hall, Ferrum VA 24088, USA

ABSTRACT

Ailanthus (*Ailanthus altissima*) is an exotic invasive tree species that is becoming an increasingly common threat to native forest communities in the eastern United States. The species spreads aggressively by both seed and sprouts and produces allelopathic chemicals that may enable its spread. We tested two chemical control methods on Ailanthus (cut stump application with an aqueous 50% formulation of triclopyr vs. basal bark application using a 20% ester formulation of triclopyr). Each application was applied to replicated groups of trees in the spring following budbreak or in the fall before leaf fall. The cut stump application resulted in 100% control of *Ailanthus* (no resprouting) in both spring and fall, while the basal bark application was >90% effective (complete crown mortality). Spring applications were slightly more effective than fall applications for basal bark treatments. While chemical control appears to be an effective method for killing *Ailanthus* trees, abundant germination from the soil seed bank surrounding controlled trees will require long-term control efforts.

INTRODUCTION

Ailanthus (*Ailanthus altissima*), also known commonly as tree-of-heaven, is a widespread invasive exotic tree species in the eastern United States. The tree is native to China and had been planted extensively as an ornamental tree in the 19th century. Ailanthus tolerates a wide variety of environmental conditions (Miller, 1990), although it is intolerant of shade (Grime,

* tfredericksen@ferrum.edu

1965). It can spread rapidly from both seeds and sprouts (Miller, 1990). Ailanthus is also produces allelopathic compounds (Mergen, 1959; Heisey, 1990; Lawrence et al. 1991) which may advance its spread into forests.

Trees store carbohydrates in the stem and roots during the dormant season and cutting hardwood tree species during the growing season compared to the dormant season has been shown to reduce sprouting vigor (Kays et al., 1988). Manual cutting of ailanthus has been shown to stimulate resprouting and increase overall stand density (Burch and Zedaker, 2003). Herbicide application to freshly cut stumps can control or prevent resprouting (Zedaker et al., 1987). Since cutting of trees is labor intensive and damage to residual vegetation or property could occur during felling, an alternative for killing ailanthus trees with felling is basal bark application (Burch and Zedaker, 2003). This control method uses an oil-penetrating formulation of a systemic herbicide applied to the bark, which translocates throughout the tree.

With either herbicide application method, the timing of application during the growing season to achieve the best control is uncertain. On one hand, manual cutting or application of herbicides to deciduous tree species during the spring may be effective because trees have recently exhausted their carbohydrate reserves after refoliation. On the other hand, fall application of herbicide treatments may provide more effective control because of movement of photosynthate in trees towards the roots for winter storage will increase the amount of a systemic herbicide translocated to the roots, increasing the likelihood of root mortality.

A study was initiated in 2003 to compare the efficacy of herbicide application method (cut stump application vs. basal bark treatment) and season of application (spring vs. fall) for controlling ailanthus on the campus of Ferrum College, Ferrum, Virginia, USA.

METHODS

Treatment applications were made to groups of 2-34 trees on the Ferrum College campus. Trees were 5-20 cm diameter at breast height (DBH). Each tree received a numbered tag attached to the stem with a galvanized nail just above ground level. Treatments were applied to groups of trees rather than individual trees because clonal relationships or grafting among stems within a group were possible. Twenty groups of ailanthus trees were identified and randomly assigned to one of four factorial combinations: spring vs. autumn application and cut stump vs. basal bark application (five replicates of each treatment combination). Cut stump application including felling trees just above the surface of the ground and applying a 50% (by volume) aqueous solution of triclopyr (Garlon 3A®) to the entire surface of the freshly cut stump using a squirt bottle. The basal bark application consisted of applying a 20% (by volume) solution of the ester-based formulation of triclopyr (Garlon 4E®) in diesel fuel directly to the basal portion of the uncut stem (approximately 1 m above ground). The solution was applied using a squirt bottle in a 20 cm wide band entirely circling the tree at approximately 1 m above ground.

Fall applications were made in early October, 2003 and spring applications were made in early June, 2004. Evaluations were made in June, 2005. Each tree was evaluated individually and scored using one of four categories based on the extent of control by either crown mortality (for basal bark treatments) or resprouting vigor (cut-stump treatments).

Crown mortality classes included: no apparent effect, minor to moderate effect (< 50% crown mortality), major effect (> 50% crown mortality), or total control (100% crown mortality). Resprouting classes included: no effect (vigorous resprouting), minor effect (some resprouting, but not vigorous), major effect (few resprouts, poor vigor), and total control (no resprouting).

Data Analysis

The number of trees in each group were tallied and the percentage of trees responding in each class was averaged and the mean group response was used in chi-square contingency analysis. Data were analyzed using the SYSTAT 10.2 (SYSTAT Software, Inc., San Jose, CA) statistical analysis program. Statistical tests were considered statistically significant at $p \leq 0.05$.

RESULTS

All treatments were highly effective in controlling ailanthus, ranging from 91-96% in crown mortality for basal bark treatments to 100% control of resprouting for cut stump treatments 12-19 months after application (Table 1). The frequency distribution of the different response classes was significantly skewed towards complete control of ailanthus ($\chi 2 = 12.4$, $p = 0.006$). The average percentage of trees per group in the complete control class did not differ between cut-stump and basal bark treatments ($\chi 2 = 0.1$, $p = 0.75$). For basal bark treatments, complete control did not differ between spring or fall applications ($\chi 2 = 0.07$, $p = 0.8$).

DISCUSSION

Ailanthus trees appear to be easy to control with the herbicides used in this study. Burch and Zedaker (2003) also observed high control rates of ailanthus using basal application, with defoliation rates of 75-100% after six weeks using a variety of herbicide-carrier systems, most of which included Garlon 4. They observed 100% defoliation using Garlon 4 in a 20% formulation with HY-Grade EC used a carrier. The timing of application for that study was in June. The lower efficacy of both June and October basal bark applications in our study may be attributed to the use of kerosene as a carrier. While kerosene is relatively inexpensive and widely available, it is not specially formulated for herbicide applications as those used in Burch and Zedaker (2003).

The somewhat lower efficacy of the fall basal application treatment in our study in the fall may be due to the date of application (early October). This date is relatively late with respect to fall foliage senescence. Optimally, fall application should occur in late August through mid-September before the onset of leaf senescence which may limit translocated of the herbicide throughout the tree. The timing of spring application is probably also important, with the best likely timing occurring just after full leaf expansion, when trees have fully exhausted carbohydrate reserves, but have not passed the point of attaining a positive photosynthetic carbon balance.

Table 1. Mean (± 1 standard error) percentage of trees in response classes of groups (n = 5) of ailanthus trees treated with cut stump application with Garlon 3A (degree of resprouting) or basal bark application with Garlon 4E (extent of crown mortality) either during the spring (early June, 2004) or fall (early October, 2003) as evaluated in June, 2005 (12 -19 months after application).

Treatment	Crown dead - no resprouting	> 50% crown mortality - few weak sprouts	< 50% crown mortality - some resprouting	No visible crown effect - vigorous resprouting
Cut stump - spring	100	0	0	0
Basal bark – spring	95.8 ± 5.3	4.2 ± 5.3	0	0
Cut stump – fall	100	0	0	0
Basal bark - fall	91 ± 5.6	9 ± 5.6	0	0

Regardless of application type of season, the use of herbicides is important for controlling ailanthus. Manual cutting of ailanthus provides only temporary control and may ultimately lead to larger infestations of this species because of its ability to vigorously resprout from cut stumps (Burch and Zedaker, 2003). The rapid growth rates of ailanthus, particularly from resprouts (Miller, 1990) may allow for rapid recovery after cutting. Even with the use of herbicides, long-term follow-up control of treated areas is probably necessary. Ailanthus has prolific seed production (Miller, 1990) and the soil seed bank will likely produce seedlings at the base of treated trees for many years and will require repeated herbicide spraying.

Controlling the spread of ailanthus will likely be a long-term endeavor in many parts of the eastern United States. Ailanthus trees appear to be easy to kill, but the species difficult to control. Established stands of the species along forest edges, roadways, and disturbed areas provide seed sources for colonization whenever disturbances produce exposed soil and high light availability favorable for establishment of ailanthus. Ailanthus colonization of the interiors of logged stands in Virginia has been observed (Carter and Fredericksen, 2007) and herbicide treatment of ailanthus at the edge of these stands at the time of cutting may help reduce the extent of invasion by this species. Post-logging control of ailanthus seedlings and saplings may also be necessary. Knapp and Canham (2000) also found that ailanthus can invade natural gaps in old growth forests in New York. If established in closed forests, Ailanthus may also be able to persist through clonal propagation (Kowarik, 1995). Control of the spread of this species should focus on the removal of seed trees and vigilant control of regeneration both along forest edges and within forests where colonization has occurred.

Acknowledgments

We acknowledge the support of Pat Burch and Dow Agrosciences for donation of herbicide formulations used in this study. Garlon 3 and Garlon 4 are trademarks of Dow Agrosciences, LLC.

References

Burch, P.L. & S.M. Zedaker. Removing the invasive tree Ailanthus altissima and restoring natural cover. Journal of Arboriculture 29:18-23.

Carter, W.K. & T.S. Fredericksen. 2007. Tree seedling and sapling density and deer browsing incidence on recently-logged and mature non-industrial private forestlands in Virginia, USA. Forest Ecology and Management. 242:671-677.

Grime, J.P. 1965. Shade tolerance in flowering plants. Nature 208:161-163.

Heisey, R.M. 1990. Evidence for allelopathy by tree-of-heaven (*Ailanthus altissima*). Journal of Chemical Ecology 16:2039-2055.

Kays, J.S., D.W. Smith, S.M. Zedaker, & R.E. Kreh. 1988. Factors affecting natural regeneration of Piedmont hardwoods. Southern Journal of Applied Forestry 11: 46-49.

Kowarik, I. 1995. Clonal growth in *Ailanthus altissima* on a natural site in West Virginia. Journal of Vegetation Science 6:853-856.

Knapp, L.B. & C.D. Canham. 2003. Invasion of an old-growth forest in New York by *Ailanthus altissima*: sapling growth and recruitment in canopy gaps. Bulletin of the Torrey Botanical Club 127:307-315.

Lawrence, J.G., A. Colwell, & O.J. Sexton. 1991. The ecological impact of allelopathy in *Ailanthus altissima* (Simaroubaceae). American Journal of Botany 78:948-958.

Mergen, F. 1959. A toxic principle in the leaves of *Ailanthus*. Botanical Gazette 121:32-36.

Miller, J.H. 1990. Ailanthus. Pp. 101-104 *In* Burns, R.M.& B.H. Honkala (Technical Coordinators). Silvics of North America: Volume 2. Hardwoods Agriculture Handbook 654, U.S. Dept. of Agriculture, Forest Service, Washington, D.C., 877 p.

Zedaker, S.M., J.B. Lewis, D.W. Smith, & R. E. Kreh. 1987. Impact of season of harvest and site quality on cut-stump treatment of Piedmont hardwoods. Southern Journal of Applied Forestry 11:46-49.

In: Invasive Species: Detection, Impact and Control
Editors: C.P. Wilcox and R.B. Turpin	ISBN 978-1-60692-252-1

Short Communication 2

DROWNING IN THE SAND: INVASION BY FORAMINIFERA

Mehmet Baki Yokeş[1,*] ***and Engin Meriç***[2,+]
[1] Halic Üniversitesi, Fen-Edebiyat Fakültesi, Moleküler Biyoloji ve Genetik Bölümü, Findikzade, 34093 Istanbul, Turkey
[2] Moda Hüseyin Bey Sokak, 15/4 Kadikoy, 34710 Istanbul, Turkey

ALIEN FORAMINIFERA IN THE MEDITERRANEAN

Invasion of the Mediterranean Sea by alien species is getting increasingly prominent every day. These alien species are probably transported in with ballast waters or by attaching themselves onto vessels. Furthermore, many species have been brought into the Mediterranean for aquaculture purposes or inadvertantly introduced from public aquariums. But, most of the alien taxa recorded in the Mediterranean are originated from Indo-Pacific, indicating that the Suez Canal is considered to be the major vector in introducing alien species into the Levantine Basin. More than five hundered alien species have been recorded so far and each year new arrivals are reported (Galil, 2008). Alien invasion is considered to be a major problem worldwide, especially if the native coastal ecosystems are polluted and disturbed by the anthropogenic activities, such as observed in the Mediterranean. The aliens can deplete food sources, change the habitat structure and environmental conditions, in which the native species cannot survive. However, the aliens with an economical value constitute an important portion of the fisheries in the Eastern Mediterranean.

The macroscopic aliens, such as algae, fishes, crustaceans and molluscs have been well studied, but until recently, very little attention has been paid to the microscopic ones, such as foraminifers. In a recent study, 34 genera and 45 species of foraminifera reported from the Mediterranean basin (Zenetos et al., 2008) (PLATES 1-5). Some of these alien foraminifer species are found in large quantities, locally more abundant than the native ones, but the

* bakiyokes@halic.edu.tr
+ barutif@istanbul.edu.tr

PLATE 1

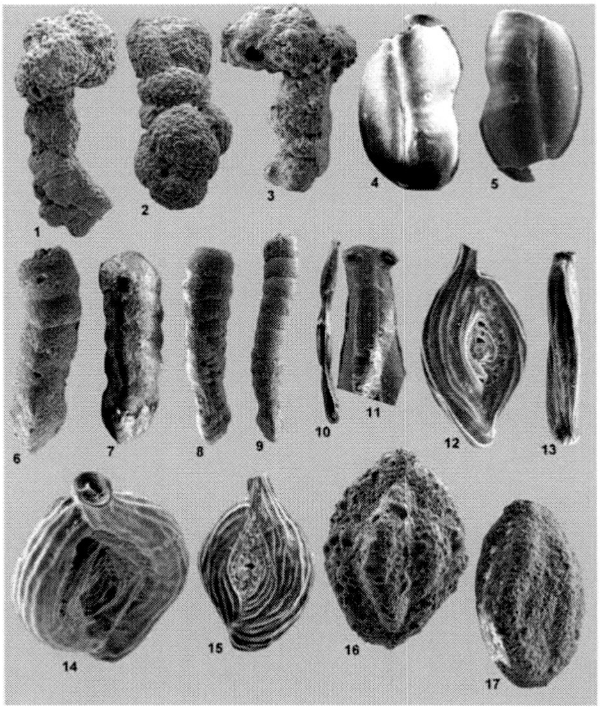

1. *Haddonia* sp. Side view, x 11.5, Kekova-Antalya, Station 84/20 m.
2. *Haddonia* sp. Side view, x 13, Kas-Antalya, Station 86/15 m.
3. *Haddonia* sp. Side view, x 11.8, Kekova-Antalya, Station 91/24 m
4. *Edentostomina cultrata* (Brady). Side view, x 46, Marmaris Bay, Station 7/29.9 m.
5. *Edentostomina cultrata* (Brady). Side view, x 29, Marmaris Bay, Station 7/29.9 m.
6. *Clavulina angularis* d'Orbigny. Side view, x 23.5, Kas-Antalya, Station 131/12 m.
7. *Clavulina angularis* d'Orbigny. Side view, x 26, Kalkan-Antalya, Station 18/6 m.
8. *Clavulina* cf. *C. multicamerata* Chapman. Side view, x 12.5, Kas-Antalya, Station 50/14 m.
9. *Clavulina* cf. *C. multicamerata* Chapman. Side view, x 15.5, Kas-Antalya, Station 34/13 m.
10. *Nodopthalmidium antillarum* (Cushman). Side view, x 28, Gulf of Iskenderun, Station 95/22 m.
11. *Nodopthalmidium antillarum* (Cushman). Side view, x 135, Gulf of Iskenderun, Station 95/22 m.
12. *Spiroloculina* cf. *S. angulata* d'Orbigny.Side view, x 30.5, Kas-Antalya, Station 88/5 m.
13. *Spiroloculina* cf. *S. angulata* d'Orbigny. Side view, x 21.5, Kas-Antalya, Station 97/12 m.
14. *Spiroloculina antillarum* d'Orbigny. Side view, x 45, Ucadalar-Antalya, Station 127/21 m.
15. *Spiroloculina antillarum* d'Orbigny. Side view, x 29, Kas-Antalya, Station 131/12 m.
16. *Schlumbergerina alveoliniformis* (Brady). Side view, x 31, Kas-Antalya, Station 84/3 m.
17. *Schlumbergerina alveoliniformis* (Brady). Side view, x 26, Kas-Antalya, Station 42/9 m.

majority of them are rare represented by 1 to 10 individuals per gram sediment. The rarely observed genera and species are *Iridia diaphana* HERON-ALLEN & EARLAND, *Haddonia sp.*, *Edentostomina cultrata* (BRADY), *Clavulina angularis* d'ORBIGNY, *C.* cf. *multicamerata* CHAPMAN, *Nodopthalmidium antillarum* (CUSHMAN), *Agglutinella arenata* (SAID), *A. compressa* EL-NAKHAL, *A. robusta* EL-NAKHAL, *A. soriformis* EL-NAKHAL, *Schlumbergerina alveoliniformis* (BRADY), *Quinqueloculina* cf. *mosharrafai* SAID, *Miliolinella* cf. *hybrida* (TERQUEM), *Pseudomassilina reticulata* (HERON-ALLEN & EARLAND), *Pyrgo denticulata* (BRADY), *Triloculina* cf. *fichteliana* d'ORBIGNY,

Articulina alticostata CUSHMAN, *Cyclorbiculina compressa* (d'ORBIGNY), *Peneroplis antillarum* d'ORBIGNY, *Borelis* sp., *Pyramidulina catesbyi* (d'ORBIGNY), *P. perversa* (SCHWAGER), *Astacolus insolithus* (SCHWAGER), *A. sublegumen* (PARR), *Favulina melosquamosa* (McCULLOCH), *Cushmanina striatopunctata* (PARKER & JONES), *Entosigmomorphina* sp., *Pulleniatina obliquiloculata* (PARKER & JONES), *Euuvigerina* sp., *Cymbaloporetta plana* (CUSHMAN), *C. squammosa* (d'ORBIGNY), *Acervulina inhaerens* SCHULTZE, *Planogypsina acervalis* (BRADY), *P. squamiformis* (CHAPMAN), *Amphistegina lessonii* d'ORBIGNY, *A. madagascariensis* (d'ORBIGNY), *Planorbulinella larvata* (PARKER & JONES), *Elphidium* cf. *charlottense* (VELLA), *E. striatopunctatum* (FICHTEL & MOLL), *Heterocyclina tuberculata* (MOBIUS), *Operculina ammonoides* (GRONOVIUS).

PLATE 2

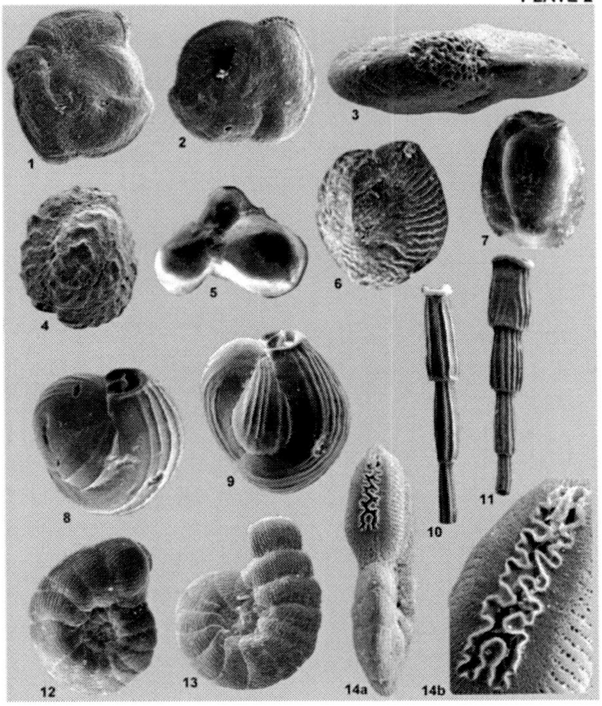

1. *Hauerina diversa* Cushman. Side view, x 29, Kas-Antalya, Station 97/12 m.
2. *Hauerina diversa* Cushman. a, side view, x 23 and b, detail view of the aperture, x 112, Kas-Antalya, Station 97/12 m.
3. *Hauerina diversa* Cushman. Edge and apertural view, x 53, Kas-Antalya, Station 97/12 m.
4. *Quinqueloculina* cf. *Q. mosharrafai* Said. Side view, x 22, Kas-Antalya, Station 81/12 m.
5. *Miliolinella* cf. *M. hybrida* (Terquem). Side view, x 26, Kas-Antalya, Station 81/24 m.
6. *Pseudomassilina reticulata* (Heron-Allen and Earland). Side view, x 35, Kalkan-Antalya, Station 18/6 m.
7. *Pyrgo denticulata* (Brady). Side view, x 21, Kas-Antalya, Station 81/15 m.
8. *Triloculina* cf. *T. fichteliana* d'Orbigny. Side view, x 54, Kas-Antalya, Station 78/24 m.
9. *Triloculina* cf. *T. fichteliana* d'Orbigny. Side view, x 54, Kas-Antalya, Station 78/24 m.
10. *Articulina alticostata* Cushman. Side view, x 19.5, Gulf of Iskenderun, Station 5/27 m.
11. *Articulina alticostata* Cushman. Side view, x 34.5, Gulf of Datça, Station 4/40 m.
12. *Peneroplis arietinus* (Batsch). Juvenile, side view, x 22, Kas-Antalya, Station 92/24 m.
13. *Peneroplis arietinus* (Batsch). Side view, x 21, Kas-Antalya, Station 92/24 m.

14. *Peneroplis arietinus* (Batsch). a, side view, x 35.5 and b, detail view of aperture, x 101, Kalkan-Antalya, Station 15/14 m.

PLATE 3

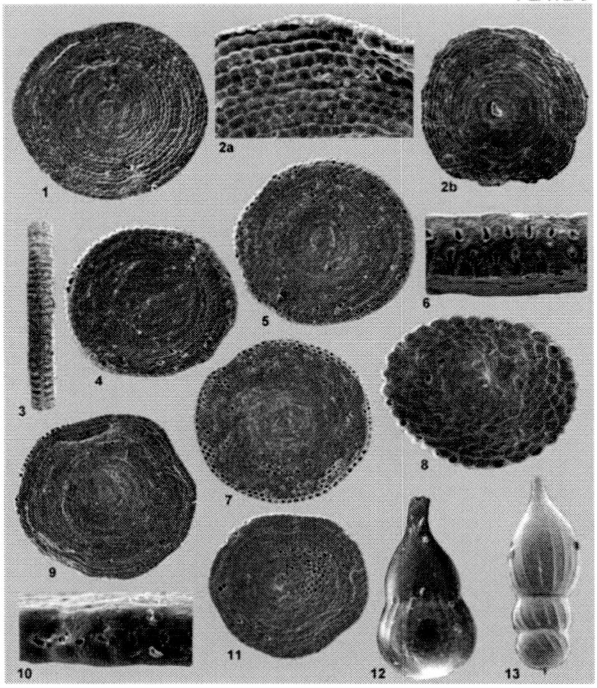

1. *Cyclorbiculina compressa* (d'Orbigny). Side view, x 12 and 8a enlargement of side view, x 35.5, Kas-Antalya, Station 63/24 m.
2. *Cyclorbiculina compressa* (d'Orbigny). a, side view, x 3.6; b, enlargement of side views, b, x 14, Kas-Antalya, Station 60/21 m.
3. *Amphisorus hemprichii* Ehrenberg. Peripheral view, x 13 Kas-Antalya, Station 99/24 m.
4. *Amphisorus hemprichii* Ehrenberg. Side view, x 12.5 Kas-Antalya, Station 99/24 m
5. *Amphisorus hemprichii* Ehrenberg. Side view, x 11.5, Kas-Antalya, Station 99/24 m.
6. *Amphisorus hemprichii* Ehrenberg. Enlargement of peripheral view, x 41 Kekova-Antalya, Station 99/24 m.
7. *Sorites orbiculus* Ehrenberg. Side view, x 9, Kekova-Antalya, Station 91/6 m.
8. *Sorites orbiculus* Ehrenberg. Side view, x 27.5, Kas-Antalya, Station 131/12 m.
9. *Sorites variabilis* Lacroix. Side view, x 9, Kas-Antalya, Station 49/14 m.
10. *Sorites variabilis* Lacroix. edge view, x 16.5, Kas-Antalya, Station 131/12 m.
11. *Sorites variabilis* Lacroix. Side view, x 13, Kas-Antalya, Station 131/12 m.
12. *Pyramidulina catesbyi* (d'Orbigny). side view, x 79, Gulf of Iskenderun, Station 95/22 m.
13. *Pyramidulina perversa* (Schwager). Side view, x 55, Gulf of Izmir, Station 11/24 m.

Besides the rarely observed species, there are some species which locally dominate the foraminifer fauna and have larger populations made up of 50 to 100 individuals per sample, such as, *Spiroloculina angulata* CUSHMAN, *S. antillarum* d'ORBIGNY, *Hauerina diversa* CUSHMAN, *Peneroplis arietinus* (BATSCH), *Amphisorus hemprichii* EHRENBERG, *Sorites orbiculus* EHRENBERG, *S. variabilis* LACROIX, *Amphistegina lobifera* LARSEN, *Heterostegina depressa* d'ORBIGNY (Hottinger, 1977; Baccaert, 1987; Loeblich & Tapan, 1988, 1994; Cimerman & Langer, 1991; Hatta & Ujiie, 1992; Hottinger et al., 1993; Yanko et al., 1993; Yanko, 1995; Avşar & Yanko, 1995; Yasinsi & Jones, 1995; Avşar, 1997; Haunold et al., 1997; Hayward et al., 1997, 1999; Langer and Hottinger, 2000; Meriç & Avşar, 2001; Avşar et al., 2001; Samir et al., 2003; Meriç et al., 2004 a, b; 2008 a, b, c).

PLATE 4

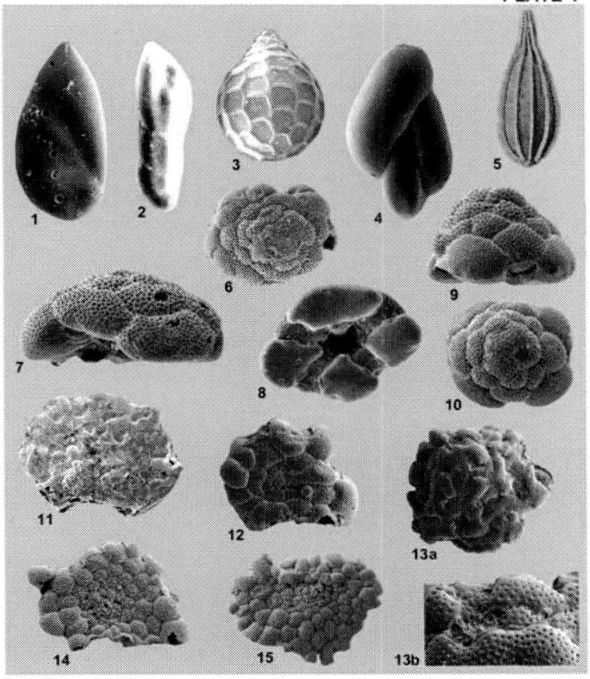

1. *Astacolus insolithus* (Schwager). Side view, x 38, Gulf of Gökova, Station 3/80.4 m.
2. *Astacolus sublegumen* (Parr). Side view, x 34, Gulf of Edremit, Station 18/29 m.
3. *Favulina melosquamosa* (McCulloch). Side view, x 89.5, Dardanelles Strait, Station 7/68 m.
4. *Entosigmomorphina* sp. a, side view, x 28.5 and b, aperture, x 108.5, Kas-Antalya, Station 131/19 m.
5. *Cushmanina striatopunctata* (Parker & Jones). Side view, x 45, Dardanelles Strait, Station 2/35 m.
6. *Cymbaloporetta plana* (Cushman). Spiral side, x 29, Ucadalar-Antalya, Station 119/24 m.
7. *Cymbaloporetta plana* (Cushman). Edge view, x 57, Kas-Antalya, Station 73/8 m.
8. *Cymbaloporetta plana* (Cushman). Umblical side, x 41.5, Kas-Antalya, Station 73/8 m.
9. *Cymbaloporetta squammosa* (d'Orbigny). Edge view, x 48, Kas-Antalya, Station 84/10 m.
10. *Cymbaloporetta squammosa* (d'Orbigny). Spiral side , x 46, Kas-Antalya, Station 84/12 m.
11. *Acervulina inhaerens* Schultze. Unattached side view, x 45, Gulf of Saros, Harmantasi locality, central part, 22.1 m.
12. *Planogypsina acervalis* (Brady). Unattached side view, x 20, Gulf of Saros, Harmantasi Locality, Line 4/40 m.
13. *Planogypsina acervalis* (Brady). Unattached side view, a, x 12 and b, enlargement part of periphery, x 46.5, Gulf of Saros, Harmantasi Locality, Line 4/40 m.
14. *Planogypsina squamiformis* (Chapman). Unattached side view, x 18, Kas-Antalya, Station 32/22 m.
15. *Planogypsina squamiformis* (Chapman). Unattached side view, x 12, Kas-Antalya, Station 55/14 m.

Foraminifera test is one of the principal sources of $CaCO_3$ in the tropical and subtropical seas and oceans (Kennett, 1982). The most productive is the genus *Amphistegina*. Twenty-three percent of the sand in the Hawaiian Islands is composed of *Amphistegina* tests (Hallock-Müller 1976). This percentage may reach as high as 90 % in certain locations in the Pacific (McKee et al., 1959; Hallock et al., 1995). The foraminifer species are neither toxic nor infectious, thus when compared to the invasive macroscopic aliens, they are considered to be as harmless as sand particles. Most of the alien foraminifer species coexist with other native benthic foraminifer species in the Mediterranean. However two of them, *Amphistegina lobifera* Larsen and *Amphisorus hemprichii* Ehrenberg are found to be exceptionally common on the Southwestern coasts of Turkey, enough to change the structure of the native habitat (Figure 1).

PLATE 5

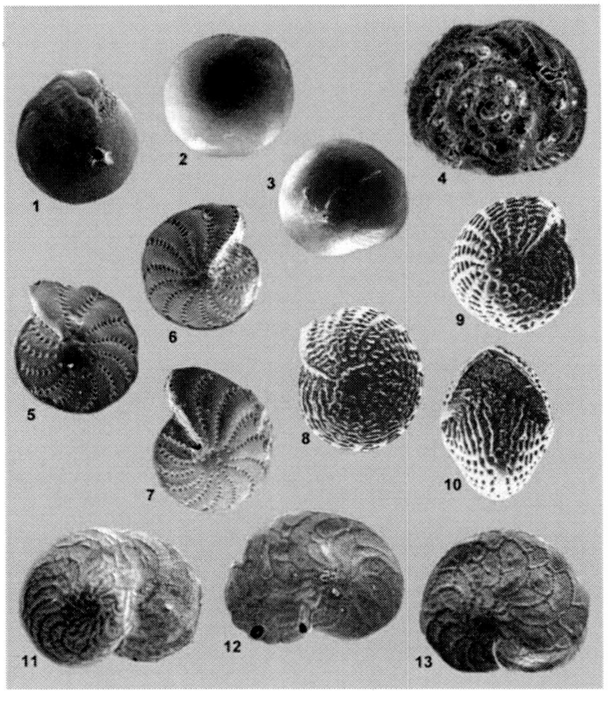

1. *Amphistegina lobifera* Larsen. Side view and aperture, x 10, Üçadalar-Antalya, Station 127/21 m.
2. *Amphistegina lobifera* Larsen. Side view, x 10.5, Ucadalar-Antalya, Station 127/21 m.
3. *Amphistegina lobifera* Larsen. Side view, x 10, Ucadalar-Antalya, Station 127/21 m.
4. *Amphistegina lobifera* Larsen. equatorial section, x 27.5, Kas-Antalya, Station 43/14 m.
5. *Elphidium charlottense* (Vella). Side view, x 26, Kekova-Antalya, Station 81/24 m.
6. *Elphidium charlottense* (Vella). Side view, x 24.5, Kekova-Antalya, Station 101/18 m.
7. *Elphidium charlottense* (Vella). Side view, x 25, Ucadalar-Antalya, Station 127/18 m.
8. *Elphidium striatopunctatum* (Fichtel and Moll). Side view, x 33, Gulf of Iskenderun, Station 70/25 m.
9. *Elphidium striatopunctatum* (Fichtel and Moll). Side view, x 33, Gulf of Iskenderun, Station 70/25 m.
10. *Elphidium striatopunctatum* (Fichtel and Moll). apertural view, x 39, Gulf of Iskenderun, Station 70/25 m.
11. *Heterostegina depressa* d'Orbigny. Side view, x 9, Besadalar-Antalya, Station 103/12 m.
12. *Heterostegina depressa* d'Orbigny. Side view, x 17.5, Ucadalar-Antalya, Station 127/18 m.
13. *Heterostegina depressa* d'Orbigny. Side view, x 18.5, Ucadalar-Antalya, Station 127/21 m.

Amphistegina lobifera Larsen

The most abundant of the alien foraminifer species in the Mediterranean Sea is *Amphistegina lobifera* Larsen. It shows a wide distribution range in the Indo-Pacific and Atlantic Oceans (Langer and Hottinger, 2000), also abundantly recorded in the Eastern

Mediterranean Basin, Israel (Langer and Hottinger, 2000; Hyams et al., 2002), Lebanon (Moncharmont Zei, 1968) Greece (Cherif, 1970; Hollaus and Hottinger, 1997) and Turkey (Avsar, 1997; Meric et al., 2002, 2004a). In Israel, *Amphistegina lobifera* is the most abundant foraminifer species found in hard substrate, reaching densities of almost 180 specimens/g (Hyams et al., 2002). Its range of distribution in Mediterranean is mainly restricted to Levantine basin, and in the Central Mediterranean it has been reported only from Libya (Blanc-Vernet et al., 1979; Crapon-De Caprona and Benier, 1985), Tunisa (Glacon, 1962) and Malta (Yokes et.al, 2007).

Figure 1. Satellite photograph of the southwestern coast of Turkey. © SeaWiFS Project, NASA/Goddard Space Flight Center and ORBIMAGE.

A. lobifera is observed almost everywhere on the Aegean and Mediterranean coasts of Turkey, and even in the Sea of Marmara (Meric et al., 2005). It forms extensively dense populations along the coasts of Antalya (SW Turkey) (Meric et al., 2002). The density of living individuals on the rocky substrate can reach 230000 - 310000 individuals/m^2 (Figure 2). The high ratio of tests in the sediment (>350 specimens/g; 0.75g tests/g) results in large amounts of sand formation, changing all the habitat type and coastal structure (Yokes and Meric, 2004).

The *A. lobifera* populations off Kas (Antalya, SW Turkey) have been followed between 2002 and 2008. The rich sediments were located SCUBA diving and core samples were obtained from different locations and depths. The core samples were cut into 2 cm thick slices, one gram of sediment was weighed from each slice and *A. lobifera* content in each sample were counted (Figure 3). The first scouts of this alien species were found at the bottom of the core samples, represented with very few individuals (Figure 4). Surprisingly, the size of this alien population had not shown much change for a certain period of time. However, at the 25th cm from the core surface, a rapid increase in population density was obvious. The cause of this abnormal population expansion is yet unknown, but it indicates that something had happened and The Kraken waked (Yokes and Meric, 2004).

Figure 2. Alive foraminifer individuals on the coasts of Turkey. A). *Amphistegina lobifera* Larsen population on *Cystoseira* sp., Bodrum - Mugla, depth: 9m, October 2002. B) Dense *A. lobifera* population on rocky substrate, Uc Adalar - Antalya, depth 6 m, 1490 alive individuals have been counted on 15.75cm^2 of sea bottom. C) *Amphisorus hemprichii* Ehrenberg population on *Halophila stipulacea* (Forsskål) Ascherson, Kas -Antalya, depth 26 m, November 2004. D) *A. hemprichii* population on *Cystoseira* sp., Kas - Antalya, depth 14m, November 2004.

The extensive abundace of these two alien foraminifera species results in the accumulation of tests at an extreme rate on the sea floor. Certain stations off Kas were monitored for six years and the deposition of the test belonging to these two species found to be 2-4.5cm/year. The thickness of the deposited test on the rocky sea bottom reaches to 60-80cm locally. Such a population density is not observed elsewhere in the Mediterranean. It is good for the sand dwelling native species, however, the native habitat sturcture in this particular area is rocky structures. Waves carry foraminifera tests to the shores where they accumulate in small bays, thus changing gravelly shores to sandy beaches. This extensive deposition of tests is creating an immense ecological problem by changing the whole habitat structure, while definitely altering the species composition of the coastal ecosystem in the long run. Thus, regardless of its biological characteristics, any alien species shoul be regarded to have a potential to cause some kind of damage to the ecosystem.

The high sea water temperature observed in this region may suggest that *Amphistegina lobifera* Larsen likes high water temperatures which may limit its distribution in the Mediterranean. According to Langer and Hottinger (2000) the occurrences of living amphisteginids are delimited by the 14°C winter isotherms. Laboratory experiments showed that *Amphistegina lobifera* Larsen ceased all movements at temperatures below 12°C (Zmiri et al., 1974). However, its presence in the northern Aegean Sea and in the eastern Sea of Marmara shows that this species can adapt to much lower temperatures, thus may be dispersed to western Mediterranean by time, the recent recording from Malta supports the idea that its invasion in the Mediterranean has not yet finished.

Figure 3. Analysis of the sediment. Close-up photography of the sediment in Kas - Antalya. A) Most of the sediment is composed of Amphistegina lobifera tests. B) Core samples were manually collected from different locations and depths by driving PVC pipes into the sand by hammering. D) High accumulation rate of foraminifera tetst results in drowning of rocky habitat in sand. The phanerogams further increase the sedimentation rate as they start growing on the test deposits.

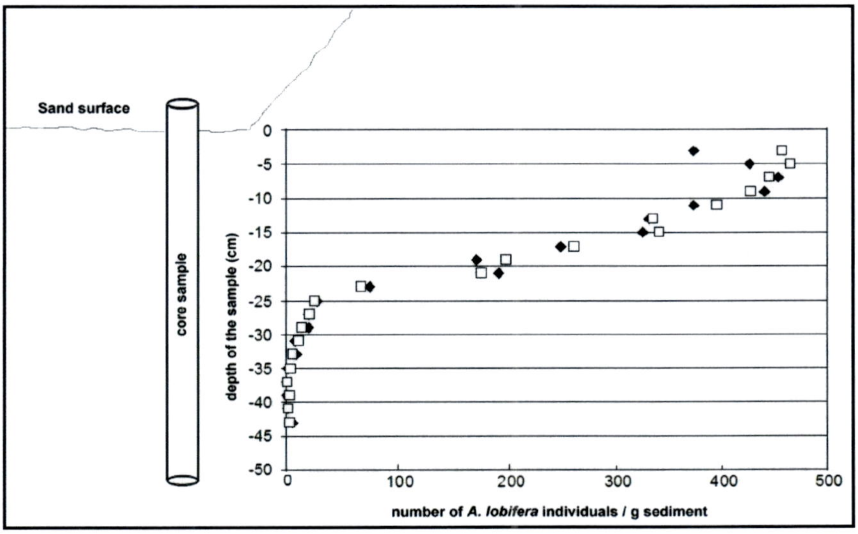

Figure 4. Analysis of the core samples. Results of two core samples from different depths show similar Amphistegina lobifera distribution pattern. ♦ core sample from 42 m, □ core sample from 46 m. Both core samples show a population expansion at the same time interval.

Amphisorus hemprichii Ehrenberg

The genus *Amphisorus* belongs to a family of calcareous larger foraminifera, Soritidae (Ehrenberg 1839), whose members are common in shallow tropical and subtropical waters. Fossil samples of *Amphisorus hemprichii* Ehrenberg were found in Creteceous rocks (Ehrenberg. 1839 and 1840). Today, it shows a wide range of distribution in the Indo-Pacific and abundantly found in the Gulf of Aqabe to the north of the Red Sea (Reiss and Hottinger, 1984; Haunold et al., 1998). The benthic foraminifer *A. hemprichii* bears algal-symbionts. This symbiotic association provides it with the necessary energy to survive in oligotrophic environments (Hallock 1999). This symbiotic relationship also promotes exceptional test growth in foraminifers by enhancing calcification (ter Kuile 1991). *A. hemprichii* populations have been first recognized in June 2002 in sediment samples from few coves around Kalkan, Kas, and in Kekova (Antalya -SW Turkey). However, in two years it covered more than 100km coastline and is still spreading in the Aegean Sea, forming uncountably dense populations on the benthos (Figure 2). Most tests are l-2mm in diameter, but a small percentage of the population is between 0.5-1 mm. In some of the stations off Kas, much larger specimens are observed, reaching 1cm in diameter. *Amphisorus hemprichii* Ehrenberg specimens are observed at a depth range of 3-24 m, but most live between 8-18 m water depth. In Aegean waters, they are found epiphitically on *Posidonia oceanica* (Linné) Delile and *Halophila stipulacea* (Forsskål) Ascherson. But, on the coasts of Kas and Kekova it is extremely abundant that they are observed not only macrophytes and phanerogams, but also on every kind of substrate, even on themselves (Figure 2).

CONCLUSION

Majority of the alien foraminifer species observed on the southwestern coast of Turkey are Indo-Pacific originated and suggested to be introduced from Red Sea. The most abundant species is *Amphistegina lobifera* Larsen. Other than *Amphistegina*, some other large Indo-Pacific foraminifer species, such as *Amphisorus hemprichii* Ehrenberg, *Sorites orbiculus* Ehrenberg and *Heterostegina depressa* d'Orbigny were found in this area. *Peneroplis arietinus* (Batsch) was abundantly observed in only this region and not reported from elsewhere in the Mediterranean. The abundance of alien foraminifer species along the southwestern coasts of Turkey is very interesting. Most of these alien benthic foraminifer species prefer warm, saline, tropical seas world-wide. Langer and Hottinger (2000) indicate that *Amphisorus hemprichii* Ehrenberg prefers 16-34 °C, *Sorites orbiculus* Ehrenberg 14-34 °C, *Amphistegina* spp. 14-34 °C and *Heterostegina depressa* d'Orbigny 19-34 °C. But the conditions in the Mediterranean and Aegean coasts of Turkey are remarkably different. The presence of Mesozoic and Cenozoic limestones with karstic characteristics in the Taurus Mountains combined with numerous cold water springs along the coastline cause salinity and temperature variations in these regions. 23 submarine springs have been located along the coastline between Kalkan and Kekova Island (Öztan et al., 2004). The differences in distribution patterns and abundances of alien species suggest the diversity of ecological characteristics in the region. It is still under debate whether such a large number of Erythrean species will exert a competitive pressure on the local biota resulting in the disappearence of native species. However there is an obvious ongoing process of destruction of the coastal ecosystem along the southwestern coasts of Turkey, where the damage is directly caused by the increased population of the Erythraen foraminifer *Amphistegina lobifera* Larsen.

It has been known that some of the alien species showed unusual population explosions in a very short time, soon after they were first recorded in the Mediterranean (Golani, 2004). A similar situation is seen in the *Amphistegina lobifera* at first glance, but the analysis of the core samples showed that the *A. lobifera* population has been stable for a while, before a rapid expansion has occured. The cause of this expansion might be a change in environmental conditions; such as nutrients, temperature or a change in the chemical composition of sea water. Further analysis of the tests for age determination and mineral composition will probably reveal the reasons of this unique event and figure out the environmental history of the research area as well as the fate of the native biota.

ACKNOWLEDGEMENTS

We would like to thank Bogazici University Underwater Sports Club members for sample collection, Bülent Arman and Mehmet Ali Demirkaya, (Türkiye Şişe ve Cam Fabrikaları AŞ, Glass Research Center) for SEM photography, Feyza Dincer for plate arrangements. This study was partially funded by WWF-Turkey.

REFERENCES

Avşar, N. and Yanko, V., 1995, Taxonomy of benthic foraminifera of Iskenderun Bay.Annual Report Avicenne (AVI CT92-0007): *Benthic foraminifera as indicator of heavy metal pollution - A new kind of biological monitoring for the Mediterranean Sea.* 233-247.

Avşar, N., 1997, Foraminifera of the Eastern Mediterranean Costline. *Ç. Ü. Yerbilimleri (Geoosound),* 31, 67-81, Adana.

Avşar, N., Meriç, E. & Ergin, M., 2001. İskenderun Körfezi bentojenik sedimentlerinin foraminifer içeriği. *H. Ü. Yerbilimleri,* 24, 97-112, Ankara.

Baccaert, J., 1987, Distribution patterns and taxonomy of benthic foraminifera in the Lizard Island Reef Complex, Northern Great Barrier Reef, Australia. *Ph.D. Thesis*, Liege, C. A. P. S. Lab. Biosédimentologie.

Blanc-Vernet, L., Clairefond, P. & Orsolini, P., 1979. Les foraminiferes. *Annales de l'Universite de Provence*, 6, 171-209.

Crapon-De Caprona, d'E. A. & Benier, C., 1985. Contribution a l'etude des Soritidae Actuels (Foraminiferes). III, Sousfamilles des Archaiasinae, Meandropsininae et Soritinae et conclusions generales. *Revue de Paleobiologie*, 4, 437-490.

Cherif, O. H., 1970. Die Miliolacea der West-Küste von Naxos (Griechenland) und ihre Lebensbereiche. *PhD. Thesis, University of Clausthal, Germany,* 175 pp.

Cimerman, F. & Langer, M. R., 1991. Mediterranean foraminifera. *Slovenska Akademija Znanosti in Umetnosti, Akademia Scientiarum et Artium Slovenica.* 118 p., 93 plts., Ljubljana.

Ehrenberg, C. G., 1839, Über die Bildung der Kreidefelsen und des Kreidemergels durch unusichtbare Organiismen. *Physikalische Abhandlungen der Königlichen Akademie der Wissenschaften zu Berlin,* 59-147.

Ehrenberg, C. G., 1840, Über noch jetzt zahlreich lebende Thierarten der Kreidebildung und den Organismus der Polythalamien. *Physikalische Abhandlungen der Königlichen Akademie der Wissenschaften zu Berlin,* 1839 (1841: separate 1840), 81-174.

Glacon G (1962) Foraminiferes des dépôts actuels des côtes de Tunisie Sud Orientale. Thkse de Doctorat UniversitC de Montpellier, France 270 pp.

Galil, B. S., 2008. Alien species in the Mediterranean Sea - which, when, where, why? *Hydrobiologia*, 606, 105-116.

Golani, D., Orsi-Relini, L., Massuti, E. and Quignard, J.-P., 2004. Dynamics of fish invasions in the Mediterranean: update of the CIESM Fish Atlas. *Rapports de la Commission Internationale pour l.Exploration Scientifique de la Mer Méditerranée*, 37, 367.

Hallock Muller, P. 1976. Sediment production by shallow-water, benthic foraminifera at selected sites around Oahu, Hawaii. *Maritime sediments, Special Publication*, 1, 263-265.

Hallock, P., Talge, H. K., Cockey, E. M., & Müller, R. G. 1995. A new disease in reef-dwelling foraminifera: implication for coastal sedimentation. *Journal of Foraminiferal Research*, 25 (2), 280-286.

Hatta, A. and Ujiie, H., 1992, Benthic foraminifera from Coral Sea between Ishigaki and Iriomote Islands. Southern Ryukyu Island Arc, northwestern Pacific. *Bulletin College of Science, University of the Ryukyus*, 54, 163-287.

Haunold, T. G., Baal, C. and Piller, W. E., 1997, Benthic foraminiferal associations in the Northern Bay of Safaga, Red Sea, Egypt. *Marine Micropaleonotology,* 29, 185-210.

Haunold, T. G., BAAL, C. and Piller, W. E., 1998, Larger Foraminifera. *Abhandlungen der Senckenbergischen Naturforschenden Gesellschaft,* 548, 155-180.

Hayward, B. W., Hollis, C. J. & Grenfell, H. R., 1997, Recent Elphidiidae (Foraminiferida) of the south-west Pacific and fossil Elphidiidae of New Zealand. *Institute of Geological and Nucleear Sciences*, Monograph 16, 166 ps.

Hayward, B. W., Grenfeel, H. R., Reid, C. M. and Hayward, K. A., 1999, Recent New Zealand shallow-water benthic foraminifera: Taxonomy, ecologic distribution, biogeography and use in paleoenvironmental assessment. *Institute of Geological & Nuclear Sciences*, Monograph 21 (New Zealand Geological Survey Paleontological Bulletin 75), 258 ps.

Hollaus, S. S. & Hottinger, L., 1997. Temperature dependance of endosymbiotic relationships? Evidence from the depth range of Mediterranean *Amphistegina lessonii* (Foraminiferida) truncated by the thermocline. *Eclogae Helvetiae,* 90, 591-597.

Hottinger, L., 1977, Distribution of larger Peneroplidae, Borelis and Nummulitidae in the gulf of Elat, Red Sea. *Utrecht Micropaleonotological Bulletin,* 15, 35-110.

Hottinger, L., Halicz, E. and Reiss, Z., 1993, Recent foraminiferida from the Gulf of Aqaba, Red Sea. Slovenska Akademija Znanosti in Umetnosti, *Academia Scientiarum et Artium Slovenica*, 179 p, 230 pls.

Hyams, O., Almogi-Labin, A. & Benjamini, C., 2002. Larger foraminifera of the southeastern Mediterranean shallow continental shelf of Israel. *Israel Journal Earth Science*, 51, 169-179.

Kennett, J. P. 1982. *Marine Geology:* Prentica-Hall, Englewood-Cliffs, New Jersey, 813 p.

Langer, M. R. and Hottinger, L., 2000, Biogeography of selected “larger” foraminifera. *Micropaleontology,* 46, Supplement 1, 105-126.

Loeblich, A. R. Jr. and Tappan, H., 1988, Foraminiferal Genera and their Classification. New York, Van Nostrand Reinhold Company, 2 vols., 1182 p.

Loeblich, A. R. Jr. and Tappan, H., 1994, Foraminifera of the Sahul Shelf and Timor Sea. *Cushman Foundation for Foraminiferal Research,* Special Publication, No: 31: 661 p.

McKee, E. D., Chronic, J. & Leopold, E. B. 1959. Sedimentary belts in the lagoon of Kapingimarangi Atoll. *American Association of Peteroleum Geologists Bulletin*, 43, 501-562.

Meriç, E. & Avşar, N., 2001. Benthic foraminiferal fauna of Gökçeada Island (Northern Aegean Sea) and its local variations. *Acta Adriatica,* 42 (1),125-149.

Meriç., E, Bergin F, Yokeş MB (2002) The Prolification of Amphistegina (Lessepsian Migrants) Population at Three-Islands (Üç Adalar, Antalya), a new Observation from the Turkish Mediterranean Coast. Proceedings of Workshop on Lessepsian Migration: 27-34.

Meriç, E.,Avşar, N. & Bergin, F., 2004a. Benthic foraminifera of Eastern Aegean Sea (Turkey) Systematics and Autoecology. Chamber of Geological Engineers of Turkey and Turkish Marine Research Foundation, Publication No: 18, 306 p., 33 plts, Istanbul.

Meriç, E., Avşar, N., Nazik, A., Eryilmaz, M. & Yücesoy-Eryilmaz, F., 2004b. Saros Körfezi'nin (Kuzey Ege Denizi) güncel bentik ve planktik foraminifer toplulukları ile çökel dağılımı. *Ç. Ü. Yerbilimleri (Geosound),* 44-45, 1-44, Adana.

Meric E, Avsar N, Nazik A, Alpar B, Yokes B, Barut IF and Unlu S (2005) Gemlik Körfezi Yüzey Çökellerinin Foraminifer, Ostrakod ve Mollusk Faunası, Foraminifer Kavkılarında Gözlenen Morfolojik Anomaliler ile Bölgenin Sedimentolojik, Hidrokimyasal ve Biokimyasal Özellikleri. Maden Tetkik Arama Dergisi 131, 21-48.

Meriç, E., Avşar, N., Nazik, A., Yokeş, M., Ergin, M., Eryilmaz, M., Eryilmaz-Yücesoy, F., GÖKASAN, E., TUR, H., AYDIN, Ş. & DİNÇER, F., 2008a Çanakkale Boğazı'nın güncel bentik foraminifer, ostrakod ve mollusk topluluğu ile çökel dağılımı. (in press).

Meriç, E., Avşar, N., Yokeş, B. & Dinçer, F., 2008 b. Alibey ve Maden adaları (Ayvalık-Balıkesir) yakın çevresi bentik foraminiferelerinin taksonomik dağılımı. (in press).

Meriç, E., Avşar, N., Nazik, A., Yokeş, M. B. & Dinçer, F., 2008c Benthic foraminifer and ostracod faunas along the Kalkan. Kas, Kekova, Beş Adalar and Üç Adalar (SW Antalya) coastline *Micropaleontology* (in press).

Moncharmont-Zei, M., 1968. I foraminiferi di alcuni campioni di fondo prelevati lungo la costa di Beirut (Libano). *Bolletino Societa Nat. Napoli,* 77, 3-34.

Öztan, M., Hamarat, S., Bayari, S., Ülkenli, H., Özyurt, N., Baştanlar, Y. and Varinlioglu, G., 2004, Kas dolayı kıyı kuşağında karstlaşmanın gelişimi: Mivini ve Altug denizaltı mağaraları. Sualtı Bilim ve Teknolojisi Toplantısı Bildiriler Kitabı, 143-149, 26-28 Kasım, 2004, Sabancı Üniversitesi, Istanbul.

Reiss, Z. and Hottinger, L., 1984, The Gulf of Aqaba, *Ecological Micropaleontology,* Ecological studies, 50, Berlin-Heidelberg. Springer-Verlag, pp. 1-354.

Samir, A. M., Abdou, H. F., Zazou. S. M. & El-Menhawey, W. H.,2003. Cluuster analysis of recent benthic foraminifera from the Northwestern Mediterranean coast of Egypt. *Revue de micropaleontologie,* 46, 111-130.

ter Kuile, B. 1991, Mechanisms for calcification and carbon cycling in algal symbiont bearing foraminifera. In: Lee, J. J. & Anderson, O. R. (ed.). Biology of Foraminifera. *Academic Press,* San Diego, p. 73-90.

Yanko, V., Avşar, N., Sanvoisin, R., Spezzaferri, S., Meriç, E. & Basso, D. 1993. Foraminiferal study: Taxonomy, distribution, density and diversity. *Benthic foraminifera as indicator of heavy metal pollution-A new kind of biological monitoring for the Mediterranean,* Task 9.

Yanko, V., 1995. Benthic foraminifera as indicators of heavy metals pollution along İsraeli coast (Cruise AVI-1, May 1993). *Benthic foraminifera as indicator of heavy metal pollution-A new kind of biological monitoring for the Mediterranean,* Task 5.

Yassini, and Jones, G., 1995, Foraminiferida and ostracoda from estuarine and shelf nvironments on the southeastern coast of Australia. *University of Wollongong Press*, 484 pp.

Yokeş M., B., Meriç, E., 2004. Expanded populations of Amphistegina lobifera from the southwestern coast of Turkey. *Fourth International Congress "Environmental Micropaleontology, Microbiology and Meiobenthology", EMMM 2004*: 232-233.

Yokeş, M., B., Meriç, E. & Avşar, N., 2007. On the presence of alien foraminifera *Amphistegina lobifera* Larsen on the coasts of the Maltese Islands. *Aquatic Invasions,* 2 (4), 439-441.

Zenetos, A., Meriç, E.,Verlaque, M., Boudouresque, C., F.,Giangrande' A, Çinar, M., E., Bilecenoğlu, M., 2008. Additions to the Checklist of Marine Alien Biota in the Mediterranean with Special Emphasis on Foraminifera. *Mediterranean Marine Sciences* 9 (1), 119-165.

Zmiri A, Kahan D, Hochstein S and Reiss Z (1974) Phototaxis and thermotaxis in some species of Amphistegina (Foraminifera). *Journal of Protozoology*, 21, 133-138.

In: Invasive Species: Detection, Impact and Control
Editors: C.P. Wilcox and R.B. Turpin
ISBN 978-1-60692-252-1

Short Communication 3

ANALYSIS OF REGENERATION OF COEXISTING INTRODUCED VERSUS NATIVE SPECIES OF PINE IN THE CANARY ISLANDS

J.R. Arévalo[1,*], A. Naranjo[2], L. Agudo[3] and M. Salas[3]
[1]Department of Ecology, Universidad de La Laguna, La Laguna 38206, Islas Canarias, Spain
[2]Department of Geography, Universidad de Las Palmas de Gran Canaria, Las Palmas de Gran Canaria 35003, Islas Canarias, Spain
[3]Aula de la Naturaleza, Universidad de Las Palmas de Gran Canaria, Las Palmas de Gran Canaria 35003, Islas Canarias, Spain

ABSTRACT

Invasive alien species can have a detrimental economic impact on human enterprises such as agriculture, grazing, forestry and tourist activities. Invasive species have been identified as one the major threats to ecosystems and biodiversity, as well as human well-being. The main objective of our study is to determine whether regeneration of the exotic *Pinus pinea* is able to compete with the regeneration of the native *P. canariensis*. The study area is located in the Natural Park of Tamadaba, 1400 m asl., in the NW of Gran Canaria island (Canary Islands). Stems and regeneration of *P. canariensis* and *P. pinea* were mapped in five randomly selected plots where both species were planted together around 45 years ago. Densities and basal areas of both species were also recorded. A monitoring of the survivorship of seedlings of both species was carried out during two years. A group of individuals of *P. canariensis* were excluded from grazing to determine the effect of grazing in the survivorship of the species.

Although the dispersal ability of *P. canariensis* was more effective, once the individuals of *P. canariensis* and *P. pinea* had been established, there was not difference in survivorship. Also, we did not find differences in survivorship for individuals excluded vs. non-excluded from grazing.

Despite the stability of the exotic species, this can change with the introduction of a dispersal vector of the seeds, a squirrel, *Atlantoxerus getulus*, which was introduced in Fuerteventura (the closest island to Gran Canaria) with an estimated population of 1

* <jarevalo@ull.es>

million of individuals. Gran Canaria is also suitable for the establishment of this exotic disperser of *Pinus pinea*.

Applying a precautionary principle, control of the species will be recommended in order to avoid future problems of invasiveness of *P. pinea*, as has been found after the introduction of the dispersal vector with another squirrel in South Africa.

Key words: exotic species, regeneration dynamics, *Atlantoxerus getulus*, *Pinus pinea*, *Pinus canariensis*, Gran Canaria, grazing.

INTRODUCTION

Invasive exotic species are causing dramatic changes in many ecological systems worldwide and there is no question that invasive species are profoundly altering many communities and ecosystems, forming the second leading cause (after habitat destruction) of species endangerment and extinction (Simberloff 2001; Gurevitch & Padilla 2004).

Invasive alien species are among the major drivers of biological change in most terrestrial ecosystems (Williamson 1996; Simberloff 2001; Gurevitch & Padilla 2004) and are especially damaging to oceanic islands (Vitousek et al. 1987; Dirnbock et al. 2003). The frequency, composition, richness and invasiveness of alien plants in introduced areas can be influenced by climate and environmental stresses, geology, land use type and landscape context, competition with natives, and natural or anthropogenic disturbances (Crawley 1987; Wilson et al. 1988; 1992; Alpert et al. 2000; Richardson et al. 2000; Pyšek et al. 2002; Rejmánek et al. 2003; 2005). Therefore, identifying the factors that influence the distribution of exotic species across the landscape is fundamental for evaluation of the present and future extent of plant invasions and for the development of eradication programs (Wace 1977; Alpert et al. 2000; Rejmánek & Pitcairn 2002). In addition, exotic plant invasions can be exploited as model systems to study ecological processes in natural areas (Richardson et al. 2000; Hierro et al. 2004).

In some specific areas and communities, exotic afforestation could also provide better conditions for the establishment of the native shade-tolerant species without threatening native forests (Geldenhuys 1996). Several studies showed that plantations of exotic species have facilitated forest succession (favoring germination of native species) in their understory on sites where disturbances prevented recolonization by native forest species (e.g. Parrotta 1995; Fimbel & Fimbel 1996; Loumetto & Huttel 1997).

In previous studies, *Pinus canariensis* demonstrated a greater ability to disperse than *Pinus pinea* (Arévalo et al. 2005). The two species showed different spatial patterns, with *P. pinea* tending toward a more aggregated spatial distribution of individuals than *P. canariensis*. Bivariate spatial relationships showed no difference from a random spatial distribution, indicating the lack of any pattern of aggregation or rejection between the species. These results indicated that *P. pinea* has not spread because it is less able to disperse (strongly barochorus) than *P. canariensis* (barochorus and anemochorus).

In this study we monitored for 24 months seedlings of *Pinus canariensis* and *P. pinea* in five stands with a canopy composed of both species. In total, more than 1500 seedlings were monitored. Some of the seedlings were protected from grazing of small vertebrates, mostly rabbits (*Oryctolagus cuniculus*) in this area to detect any differences in mortality due to that in function of the species. The two hypotheses we are testing are: survivorship of seedling of the native species is higher due to a better adaptation to the environment; and grazing become an important control of density for both species of pines.

These results could provide valuable information for the development of plantation and restoration plans. A more complete understanding of these processes of coexistance of both species and their temporal dynamics will aid in the development of plantation systems that will provide a more successful environment for seedling establishment.

MATERIAL AND METHODS

Study Site

The study was conducted on the north slope of the Tamadaba Natural Park, Gran Canaria (28° 19' N, 16° 34' W), Canary Islands, Spain (Fig. 1). The park comprises 2,000 ha, some areas of which have been reforested with pine species including *Pinus canariensis*, *Pinus nigra*, *Pinus sylvestris* and *Pinus pinea*. *P. nigra* and *P. sylvestris* were used in small areas (just a few hectares) and show poor regeneration. *P. pinea* was the exotic species most widely used in reforestation, with around 500 ha planted together with *Pinus canariensis* from 1950 to 1955 (Pérez et al. 1994). The plots we selected for analysis were located in these areas.

The annual precipitation of the park is 600 mm but fog drip can supplement inputs (Kämmer 1974). The mean annual temperature is 16.8 °C. Frost events may occur a few days a year at higher altitude, but not in the study area. Because all the plots differ by less than 100 m in altitude, differences in average, maximum and minimum temperature are typically less than 1° C. Soils at the study site have been classified as per Haplumbrept and Xerochrept (Rodríguez 2000). Fires were formerly frequent within the Park but, as the result of increased management over the past 50 years, fire intervals have increased (Arévalo et al. 2001).

The dominant species is generally *P. canariensis*, although in some areas (including those selected for this study) *P. pinea* is now dominant. The understory, regardless of the dominant canopy species, is dominated by *Cistus symphytifolius*, *Bystropogon origanifolius*, *Erica arborea*, *Chamaecytisus proliferus*, *Micromeria benthamii*, *Polycarpaea aristata*, *Sonchus acaulis* and *Neotinea maculata* (Pérez et al. 1994). Nomenclature follows Izquierdo et al. (2001).

Design of the Experiment

In May 2006, we randomly located five rectangular plots in the Natural Park of Tamadaba where P. *pinea* was planted (Table I). The plots were of different sizes due to the features of the landscape, such as roads and ravines. In each plot we measured altitude and slope, and estimated canopy cover of the stand using a convex spherical densiometer (Lemmon 1957). We also visually estimated rock, bare soil, and litter cover within each complete plot on a scale of 1 to 10.

We defined trees as stems of at least 2.5 cm dbh and seedlings as stems taller than 10 cm originating from seed. Previous studies recommended these classifications in concordance with the physiognomy and phenology of these species (Salas 1994). We mapped all trees in the plots to an accuracy of 0.05 m. For the seedlings we located a transect 10 x 50 m in the middle of the plots in which all the individuals of this category were mapped. The transect was subdivided in five 10 x 10 m subplots and numbered from 1 to 5. In the subplots 2 and 4, all the pines were excluded from grazing with a protective circular metal fence 30 cm in diameter, avoiding grazing from rabbits, the only vertebrate grazer in the area.

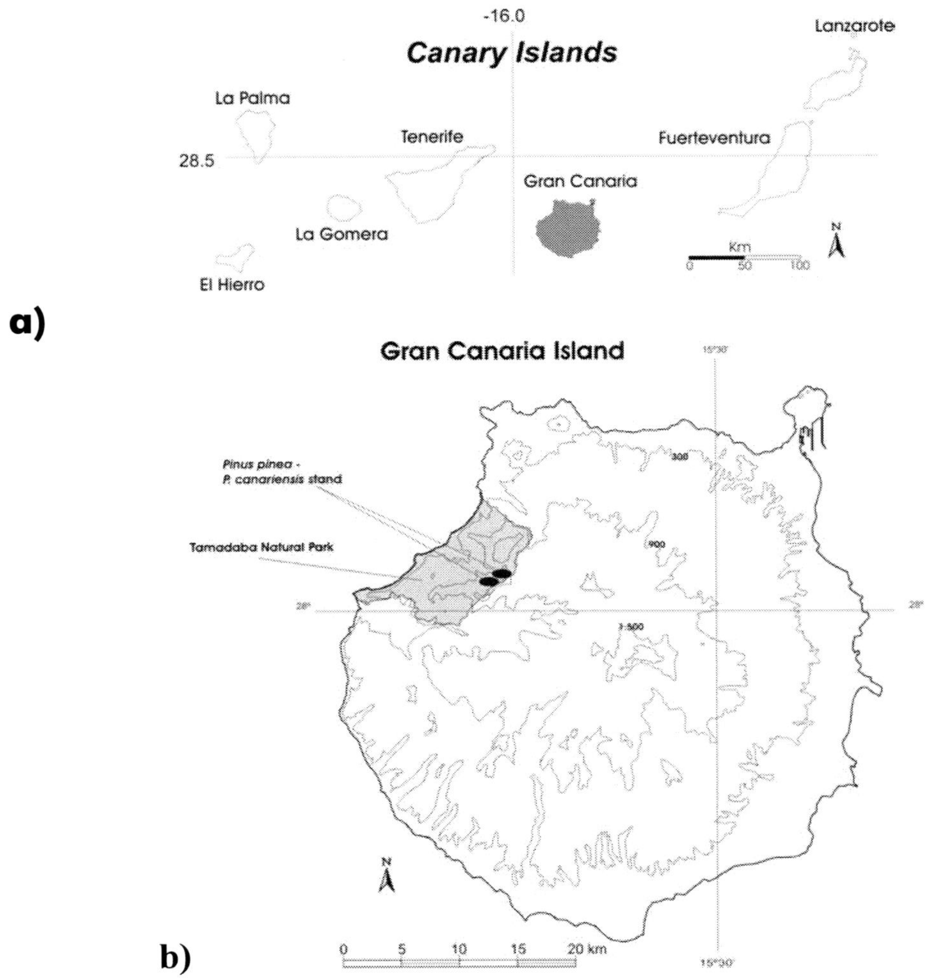

Figure 1. a) Canary Islands archipelago and b) Gran Canaria Island indicating in grey the Natural Park of Tamadaba and in black the location of the study site (geographical coordinates are indicated).

Statistical Analysis

We compared the mean percentage of survivorship at the end of the period between *Pinus canariensis* protected and unprotected from grazing in the five plots and *P. canariensis* and *P. pinea* mean percentage of survivorship with a random pair t-test ($p > 0.05$, n=5, 1000 iterations; Edgington 1985). Non-normal errors made this test suitable.

Table 1. General abiotic information of the plots

	Size m^2	Slope Sexagesimal °	Altitude m	Cover class * Rock	Soil	litter	Shrub cover class *	Canopy cover (%)
plot 1	1600	7	1205	4	3	9	6	41
plot 2	1500	21	1200	3	4	9	7	40
plot 3	1500	12	1250	5	3	9	8	60
plot 4	2000	25	1230	5	4	9	8	50
plot 5	2500	21	1140	5	6	8	7	42

(*) Cover of the different parameters estimate visually in the total plot with the following cover classes: cover classes: 1: traces, 2: <1% of cover in the plot, 3: 1-2%, 4: 2-5%, 5: 5-10%, 6: 10-25%, 7: 25-50%, 8: 50-75%, 9: >75%, 10: 100%.

The percentages of basal area and density of trees of each species (*Pinus canariensis* and *P. pinea*) were compared with the percentage of regeneration of the same species in each plot using the non-parametric Spearman Rank correlation coefficient and tested for significance ($p < 0.05$). Basic statistical methods followed Zar (1984) and were implemented using the SPSS statistical package (SPSS 1986).

RESULTS

The plots presented similar environmental characteristics, all having a canopy cover of 40-60% and an understory cover class between 6 and 8. Slope was one of the variables that offered more differences among all the plots (Table 1).

Forest stand characteristics showed more variability between plots. Measured by basal area, *P. pinea* was dominant in plots 2 and 5, while *P. canariensis* was dominant in the remaining plots (Table 2). Measured by density, *P. pinea* dominated only plot 5, while the rest of the plots were highly dominated by *P. canariensis*. Another tree species appeared in plot 4, *Erica arborea*, although its abundance was less than 1.6% of the total basal area of the plot.

Table 2. Trees (individuals of at least 2.5 cm dbh) basal area (PC: *Pinus canariensis*, PP: *Pinus pinea*, EA: *Erica arborea*) and density

	Tree (m^2/ha)			*Density (ind/ha)*		
	PC	*PP*	*EA*	*PC*	*PP*	*EA*
Plot 1	12.51	3.91	0.00	656.25	25.00	0.00
Plot 2	3.85	10.86	0.00	253.08	153.18	0.00
Plot 3	14.56	4.04	0.00	506.16	59.94	0.00
Plot 4	10.13	1.94	0.20	120.00	105.00	40.00
Plot 5	3.15	17.88	0.00	24.00	152.00	0.00

At the beginning of the study, Plot 4 had the highest number of individuals of *Pinus canariensis*, 768, while plot 3 presented the highest values for *P. pinea*, 40. At the end of the two year period, these plots were still the most abundant plots for both species respectively. The highest mortality was found in plot 2 for *P.canariensis*, and plot 3 for *P. pinea* (Annex I).

The percentage of survivorship after two years of monitoring of *Pinus canariensis* excluded from grazing was lower than the *P. canariensis* non-excluded (Fig. 2a), although differences were not significant ($t=2.807$, $n=5$, $p=0.441$). When we compared the percentage of survivorship among *P.canarienis* vs. *Pinus pinea*, the mean survivorship of *P.pinea* were higher, but the differences were not significant ($t=1.067$, $n=5$, $p=0.638$).

Annex I.- Monitoring of the individuals in five different periods, indicating the number of individuals that survive since the first sampling.

a) *Pinus canariensis* excluded from grazing

	Plot1	Plot2	Plot3	Plot4	Plot5
Feb-0	2	12	10	89	6
Jun-0(	2	12	10	85	5
Dec-0	2	11	10	82	4
Jun-0'	2	11	10	82	4
Feb-0	2	11	10	82	4

b) *Pinus canariensis* non excluded

	Plot1	Plot2	Plot3	Plot4	Plot5
Feb-06	52	54	184	598	33
Jun-06	52	53	184	584	33
Dec-06	52	49	177	546	32
Jun-07	52	49	177	548	32
Feb-08	52	49	175	547	32

c) *Pinus pinea* non excluded

	Plot1	Plot2	Plot3	Plot4	Plot5
Feb-0	2	7	40	23	5
Jun-0(	2	7	40	23	5
Dec-0	2	7	36	22	5
Jun-0'	2	7	34	22	5
Feb-0	2	7	34	22	5

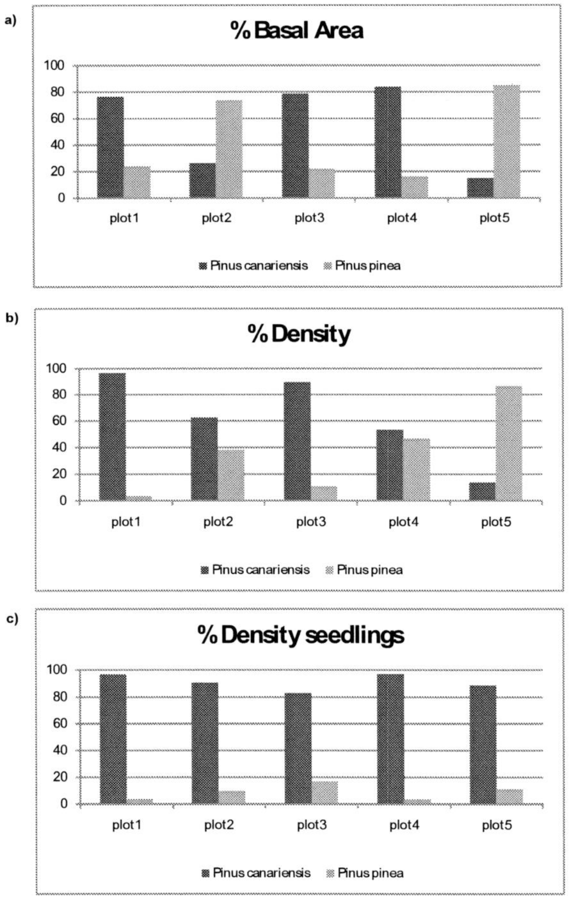

Figure 2. a) Percentage of basal areas for *Pinus canariensis* and *P. pinea* in the five plots, b) percentage of density of both species and c) percentage of density for the seedlings class.

When the percentage of regeneration density (in the last sampling period) in each plot for *Pinus canariensis* and *P. pinea* was correlated with the percentage of basal area of trees of its own species (Fig. 2a and 2c), we found no significant relation among these values ($p=0.400$ in both cases, $n=5$, $p>0.05$). When the regeneration density was correlated with the percentage of density of the trees (Fig. 2a and 2b), this relationship was also non-significant ($p=0.476$ in both cases, $n=5$, $p>0.05$).

Discussion

Because of their economic importance, pine species have been introduced out of their natural range of distribution, in some cases spreading from the areas where planted to other natural areas. *Pinus canariensis* and *P. pinea* are both invasive species in South Africa (Richardson et al. 1994) and also both of them are phylogenetically very close (Liston et al. 1999; Grotkopp et al. 2004). Therefore, the analysis of the regeneration of both species in an area where *P. canariensis* is native can reveal competitive process among similar species of pine as well as indicate the invasive character of *P. pinea*.

One of the main differences among both species of pine is the seed, Pinus *canariensis* has small seeds and a little wind can provide long distance dispersion, but in the case of *P. pinea* the seeds are large disperse mainly through barochory (dispersion by gravity providing short distances from the parent individual) on the Gran Canaria Island (Arévalo et al. 2005). There is no information about any dispersal by animals, although in other areas of the world it is dispersed by squirrels and jays (Grotkopp et al. 2004) becoming very invasive as in South Africa (Richardson 1989; Rouget et al. 2001). However, a recent invasion of a species of African squirrel, the barbary ground squirrel (*Atlantoxerus getulus*) in Fuerteventura Island, the most eastern island of the archipelago, may be able to alter this stable situation of the population of *P. pinea*. The island of Gran Canaria has been determined through several models to be one of the most suitable islands to be invaded by this species, which in Fuerteventura has been estimated to have a density between 100-200 millions of individuals (López-Darias et al., 2008).

The invader characteristics of *Pinus canariensis* can be considered stronger than the invader characteristics of *P. pinea*, as it has been demonstrated in studies based on the structure of the seed and the dispersion abilities of the species (Grotkopp et al. 2002; Réjmanek and Richardson 1996). Our results have demonstrated, that even in areas where the canopy of *P. pinea* was dominant, the regeneration were dominated by *P. canariensis* and we did not find a significant relationship among the regeneration composition and the canopy composition.

However, once the plant established, we did not find different in species survivorship in a period of 2 years (Fig. 3a), indicating that once established, and the competitive ability become more similar. We also were unable to determine a significant effect of small herbivores in *P. canariensis,* although this effect has been demonstrated important for survivorship in other species of pines (Bergman et al. 2005).

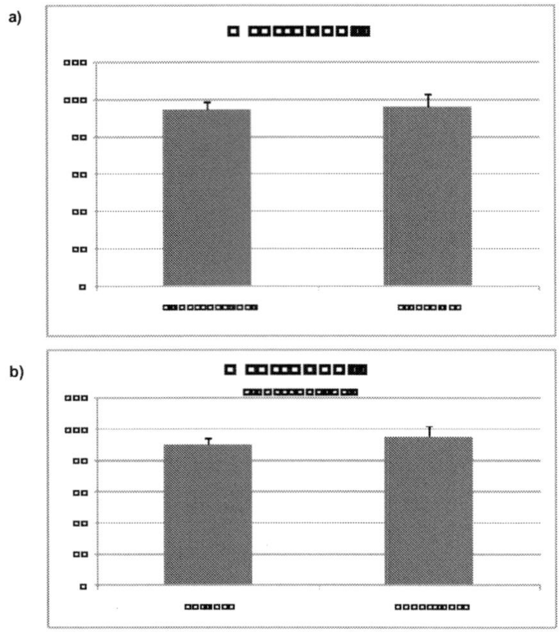

Figure 3. a) Mean percentage of survivorship for *Pinus canariensis* and *P. pinea* in the five plots and b) Mean percentage of survivorship for *P. canariensis* excluded from grazing vs. non exclude.

In previous studies were suggested that the status of *P. pinea* cannot be considered at present time dangerous from an invasive point of view. However, we suggest the necessity to control de population of the exotic pine favoring the establishment of the native one (Arévalo et al. 2005) to prevent future changes, specially knowing that the main control of the species is through the limited dispersion, and because of the broad ecological capacity of the Barbary ground squirrel and the presence of large favorable climatic areas in Gran Canaria. We also agree that in addition of the control of the pine, a management strategy is urgently needed to prevent the expansion of *A. getulus* through Gran Canaria.

Control of this species will require ecological restoration in order to favor the establishment of native vegetation that has been largely disturbed by past management practices. These recommendations, based on the ecology and biology of the species, are also in accordance with the laws of the Canarian Government, which require exotic species to be controlled and if possible, eliminated. As Walters (1986) indicated, experimentation and quantification through monitoring provide the best adaptive management approach.

ACKNOWLEGMENTS

This work is part of the program of study of exotic species that is carrying-out the Invasive Species: Interinsular Research Group (EIGI) of the University of La Laguna and University of Las Palmas de Gran Canaria. We thank the "Consejería de Educación del Gobierno de Canarias" that provided funding through its Research Support Program

(PIO42004/096). We thank the students of Geography of University of Las Palmas de Gran Canaria for their assistant in the sampling. We are also grateful for the collaboration of the fine staff of the Cabildo de Gran Canaria (especially Mr. Carlos Velazquez and Mr. Eugenio Castro), who offered the use of their facilities in the Park. We thank Heather Adams (Universidad de Las Palmas de Gran Canaria) for revising the English edition.

References

Alpert, P., Bone, E. & Holzapfel, C. (2000). Invasiveness, invasibility and the role of environmental stress in the spread of non-native plants. *Perspectives in Plant Ecology, Evolution and Systematics 3*, 52-66.

Arévalo, J.R., Fernández-Palacios, J.M., Jiménez M.J. & Gil, P. (2001). The effect of fire in the understory of two reforested stands of Pinus canariensis. Tenerife. Canary Islands. *Forest Ecology and Management 148*, 21-29

Arévalo, J.R., Naranjo, A. & Salas, M. (2005). Regeneration in a mixed stand of native Pinus canariensis and introduced Pinus pinea species. *Acta Oecologica 28*, 87-94.

Bergman, M., Iason, G.R. & Hester, A,J. (2005). Feeding patterns by roe deer and rabbits on pine, willow and birch in relation to spatial arrangement. *Oikos 109*, 513-520.

Crawley, M.J. (1987). What makes a community invasible? In Gray, A.J., Crawley, M.J. & Edwards, P.J. (Eds). *Colonization, succession and stability*, Blackwell, Oxford. pp. 429-454.

Dirnbock, T., Dullinger, S. & Grabherr, G. (2003). A regional impact assessment of climate and land-use change on alpine vegetation. *Journal of Biogeography 30*, 401-417.

Edgington, E. S. (1995). *Randomization Tests*. Marcel Dekker, Inc. New York. 409 pp.

Fimbel, A.R. & C.C. Fimbel. (1996). The role of exotic conifer plantations in rehabilitating degraded tropical forest lands: A case study from the Kivale Forest in Uganda. *Forest Ecology and Management 81*, 215-226.

Grotkopp, E., Rejmánek, M., Sanderson, M.J. & Rost, T.L. (2004). Evolution of genome size in pines (Pinus) and its life-history correlates: supertree analyses. *Evolution 58*, 1705-1729.

Grotkopp, E., Réjmanek, M. & Rost, T.L. (2002). Toward a causal explanation of plant invasiveness: seedling growth and life-history strategies of 29 pine (Pinus) species. *American Naturalist 159*, 396-419.

Geldenhuys, C.J. (1996). The Blakwood Group system: its relevance for sustainable forest management in the southern Cape. *S. Afr. For. J. 178*, 15-24.

Gurevitch, J. & Padilla D.K. (2004). Are invasive species a major cause of extinctions? *Trends Ecol. Evol. 19*, 470–474.

Hierro, J.L., Maron, J.L. & Callaway, R.M. (2005). A biogeographical approach to plant invasions: the importance of studying exotics in their introduced and native range. *Journal of Ecology 93*, 5-15.

Izquierdo, I., Martín J.L., Zurita, N. & Arechavaleta M. (2001) *Lista de especies silvestres de Canarias (hongos, plantas y animales terrestres)*. Consejería de Política Territorial y Medio Ambiente Gobierno de Canarias.

Kämmer, F. (1974). Klima und vegetation auf Tenerife, besonders in Hinblick auf den Nebelniederschlag. *Scripta Geobot. 7*, 1-78.

Lemmon, P. E. (1957). A new instrument for measuring forest overstory density. *Journal of Forestry 55*, 667-668.

Liston, A., Robinson, W.A., Piñero, D. & Alvarez-Buylla, E.R. (1999). Phylogenetics of Pinus (Pinacea) bases on nuclear ribosomal DNA internal transcribe spacer region sequences. *Molecular Phylogenetics and Evolution 11*, 95-109.

López-Darias M., Lobo, J.M. & Gouat, P. (2008). Predicting potential distributions of invasive species: the exotic Barbary ground squirrel in the Canarian archipelago and the West Mediterranean region. *Biological Invasions 10*, 1027-1040.

Loumeto, J.J. & C. Huttel. (1997). Understory vegetation in fast growing tree plantations on savanna soils in Congo. *Forest Ecology and Management 99*, 65-81.

Parrotta, J.A. (1995). Influence of overstory composition on understory colonization by native species in plantations on a degraded tropical site. *Journal of Vegetation Science 6*, 627-636.

Pérez, P. L., Pascual, M., Rodríguez, O., Acebes, J.R., del Arco, M. & Wildpret, W. 1994. *Atlas cartográfico de los pinares canarios IV: Gran Canaria y plantaciones de Fuerteventura y Lanzarote*. Viceconsejería de Medio Ambiente. S/C de Tenerife.

Pyšek, P., Jarošík, V. & Kučera, T. (2002). Patterns of invasion in temperate nature reserves. *Biological Conservation 104*, 13-24.

Réjmanek, M. & Richardson, D. (1996). What attributes make some plant species more invasive? *Ecology 77*, 1655-1661.

Rejmánek, M., Richardson, D.M., Higgins, S.I., Pitcairn, M. & Grotkopp, E. (2003). Plant invasion ecology: State of the art. Invasive alien species: looking for solutions. In: H.A. Mooney, Mack, R.N., McNeely, J.A., Neville, L., Schei, P.J. & Waage J. (Eds.). *Invasive alien species: searching for solutions*, Island Press, Washington, DC, USA, pp. 104–61.

Rejmánek, M., Richardson, D.M. & Pyšek, P. (2005). Plant invasions and invasibility of plant communities. In: Van der Maarel E. (Ed.), *Vegetation ecology*, Blackwell Science, Oxford. pp. 332–355.

Rejmánek, M. & Pitcairn, M.J. (2002). When is eradication of exotic pest plants a realistic goal? In: Veitch C.R. & Clout M. N. (eds.), *Turning the tide: The eradication of invasive species*, ,IUCN, Gland and Cambridge. p. 249–253

Richardson, D.M. (1989). *The ecology of invasions by Pinus (Pinaceae) and Hakea (Proteaceae) species, with special emphasis on patterns, processes and consequences of invasion in mountain fynbos of the southwestern Cape Province, South Africa*. PhD. diss. University of Cape Town, Cape Town.

Richardson, D.M., Williams, P.A. & Hobbs, R.J. (1994). Pine invasions in the southern hemisphere: determinants of spread and invasibility. *Journal of Biogeography 21*, 511-527.

Richardson, D.M., Pyšek, P., Rejmánek, M., Barbour, M.G., Panetta, F.D. & West, C.J., (2000). Naturalization and invasion of alien plants: concepts and definitions. *Diversity and Distributions 6*, 93–107.

Rodríguez, A. & Mora, J.L. (2000). Los suelos. In: Morales, G. and Pérez, R. (Eds.). *Gran atlas temático de Canarias*. Interinsular Canaria. S/C de Tenerife. pp. 107-120.

Rouget, M., Richardson, D.M., Milton, S.J., Polakow, D. (2001). Predicting invasion dynamics of four alien Pinus species in a highly fragmented semi-arid shrubland in South Africa. Plant Ecology 152, 79-92.

Salas, M. (1994). Estudio *del Área Potencial y Serie de los Pinares Grancanarios (Islas Canarias-España)*. PhD diss. Universidad de La Laguna. La Laguna.

SPSS. (1986). SPSS/PC+ V.6.0. Base manual. SPSS Inc., Chicago, IL.

Simberloff, D. (2001). Biological Invasions – How are they affecting us, and what can we do about them? *Western North American Naturalist 61*, 308-315.

Vitousek, P. M., Walker, L. R., Whitaker, L. D., Mueller Dombois, D. & Matson, P. A. (1987). Biological invasion by *Myrica faya* alters ecosystem development in Hawaii. *Science 238,* 802-4.

Wace, N. (1977). Assessment of dispersal of plant species: The car-borne flora in Canberra. *Proceedings of the Ecological Society of Australia 10*, 167-186.

Walters, C. (1986). *Adaptive management of removable resources*. Macmillan. New York.

Williamson, M. (1996). *Biological Invasions*. Chapman & Hall, London.

Wilson, J.B., Hubbard, J.C.E. & Rapson, G.L. (1988). A comparison of the realized niche relations of species in New Zealand and Britain. *Oecologia 76*, 106-110.

Zar, J.H. (1984) *Biostatistical Analysis*. 2nd ed. Prentice-Hall, Englewood Cliffs, NJ.

In: Invasive Species: Detection, Impact and Control
Editors: C.P. Wilcox and R.B. Turpin
ISBN 978-1-60692-252-1

Chapter 1

SPARTINA ALTERNIFLOR: A REVIEW OF ITS STATUS, DYNAMICS AND MANAGEMENT

Zifa Deng[1,2], Shuqing An[2], Zhongsheng Wang[2], Changfang Zhou[2] and Yingbiao Zhi[2]

[1] School of Life Science, Nantong University, Nantong, 226007, P. R. China
[2] Institute of Wetland Ecology and School of Life Science, Nanjing University, Nanjing 210093, P. R. China

ABSTRACT

Spartina alterniflora Loisel., a native to the Atlantic and Gulf coasts of North America, had been deliberately introduced into many countries for the control of coastal erosion and land claim. But now, the species has extensively dispersed, and even broken out in some non-native habitats. Due to Allee effect, inbreeding depression, rapid adaptation and evolution occur in the process of invasion and natural dispersal of *Spartina alterniflora*, it has become a model plant for studying biological invasion from both ecological and genetic perspectives. The previous researches showed that powerful ability of hybridization and introgression has been a genetic basic, superior reproductive capability has been the sources and strong ability of anti-stress and adaptability has been an ecological and physiologic basic for *S. alterniflora* invasion and expansion, respectively. In most invasive habitats, the expansion of *S. alterniflora,* based on intentionally transplants, indicated the mode of point dispersal. The episodic and continuous dispersal pattern of seeds has been playing an important role for maintain, recruitment and outbreak of *S. alterniflora* population. Meanwhile, consecutive expansion of *S. alterniflora* populations was ensured by the trait of potently clonal growth. Therefore, prevention of seed production in all designated areas is required to help contain this species and prevent its further spread. At the same time, although it has been proven to be very difficult, expensive and even impossible to eradicate the species, the integrated strategies with exploring the economical value and other restraining growth of *S. alterniflora* methods should be adopt to manage and alleviate the negative impacts the biological invasion.

Key words: biological invasion, Allee effect, point dispersal, ecological control

1. INTRODUCTION

Spartina alterniflora Loisel. is native to the Atlantic and Gulf coasts of North America and often dominate intertidal mudflats of salt marshes and estuaries from Newfoundland, Canada to Florida and Texas, America, and the Republic of Surinam (Daehler & Strong, 1996). Over a century, the species invaded many countries all over the world through accidental process (ie via ships' ballast water) or deliberate introduction for the control of coastal erosion and land claim (Deng et al., 2006). The extensive rhizome and root networks and upright stems have made it an ideal agent for trapping and stabilising unconsolidated estuarine muds and balancing the erosion of muddy coastal shore (Delaune et al., 1978; Mendelssohn & Kuhn, 2004). Its rapid rate of growth, high fecundity and aggressive colonisation made it very successful, displacing in some places the natural vegetation and occupying and creating new habitats (Sanchez et al., 2001; Zhang et al., 2004). Furthermore, *S. alterniflora*, as an ecological engineer, cause several alterations in the hydrology and food webs of invaded habitats that are detrimental to native wildlife and commercial uses (Qian & Ma, 1995; Zhu et al., 2004). Thus, It has been changed from an ecologically engineering species to a notorious invader. As a successful invasive species, the researches about the invasive mechanism and control strategies of *S. alterniflora* have been a hotspot.

In some places with serious erosion, the species is regarded as an important species for facilitating effective coastal defence, land-claim and even for the provision of grazing land. However, although originally considered useful and planted for coastal protection and landclaim projects around the world, views on its use and management began to change when *S. alterniflora* began to spread successfully into areas of conservation value, covering large areas of mudflats with extensive mono-specific swards. As a result, *S. alterniflora* is now possibly one of the most controversial species worldwide. Due to its biology, the habitat that it occupies, its uses and management, opinions of scientists and managers are diverse, varying almost as much as the regional uses that are made of the plant. The views of research and management practitioners on the species continue to vary in its value and management.

There is extensive literature on the subject of the biology of this non-native species, particularly in relation to its implications for the nature conservation interest of estuaries and coastal wetland. Management therefore needs to take into account the population dynamics and ecological behavior of this species, which are not as yet fully understood. The assessment presented thus includes: (1) Biological characteristics; (2) ecological superiority; (3) functioal evaluation and management .

2. BIOLOGICAL CHARACTERISTICS OF *SPARTINA ALTERNIFLORA*

2.1 Description of the Species

Spartina alterniflora Loisel., commonly known as smooth cordgrass, or Atlantic cordgrass, or saltmarsh cordgrass, is a tall clumping perennial grass with dense rhizomatous networks. The stems are hollow and grow up to 3 m in height. The leaves have flat blades which grows up to 90 cm in long, 1.5 to 2.0 cm in wide that are sometimes rolled at the tip. Flowering occurs June to November. The flowers occur in erect wands (panicles) 10 to 45 cm in long, each composed of 5 to 30 individual flowering spikes. The branch axes are triangular, with two rows of sessile, overlapping spikelets on the lower side. Spikelets are strongly flattened, keeled, and lanceolate, 8-15 mm long, with one floret and a pair of straight, unequal glumes. Spikelets detach from below the glumes. Glume and lemma keels are glabrous to sparsely covered with long, soft hairs and the ligule is hairy (Hitchcock et al. 1969; Xu, 1998). The root system of *S. alterniflora* is very developing and distributes to 30 cm, even to 1 m underground.

2.2 Genus *Spartina*

S. alterniflora belongs to the genus *Spartina*, in the Poaceae family, and the subfamily Chloridoideae which represents a well-supported mono-phyletic lineage within Poaceae. *Spartina* is a relatively small genus consisting of approximately 15 species, and no-hybrid 12 species (Wang et al., 2006). They are *S. alterniflora* Loisel., *S. argentinensis* Parodi (=*S. spartinae* Trin. (Mobberley, 1956)), *S. arundinacea* (Thouars) Carmich., *S. bakeri* Merr., S. ciliata Brongn., *S. cynosuroides (*L.) Roth, *S. densiflora* Brongn., *S. foliosa* Trin., *S. gracilis* Trin., *S. maritima* (Curtis) Fern., *S. patens* (Aiton) Muhl., *S. pectinata* Bosc ex Link, respectively. Three hybrids are *S.* × *townsendii* H. & J. Groves, *S. anglica* C. E. Hubbard and *S.* × *neyrautii* Foucaud respectively (Wang et al., 2006). According to the publications of government, there are two varieties of *S. alterniflora* species, *S. alterniflora* Loisel. var. glabra (Muhl. ex Bigelow) Fern. and *S. alterniflora* Loisel. var. pilosa (Merr.) Fern. In addition, in North America, *S. alterniflora* has two major growth forms, namely "tall form" and "short form". The tall form has a height of greater than 45 cm and the short form has a height of less than 45 cm (Valiela et al., 1978). In China, *Spartina alterniflora* presented three ecotypes, and there were significant differences among morphological and ecological parameters (Chen et al., 1992).

Most Spartina species are native to the New World (Mobberley, 1956), and only four taxa originate from Western Europe: *Spartina maritima* (Curtis) Fernald, *Spartina anglica* C.E. Hubb., *Spartina x townsendii* H. Groves & J. Groves, and *Spartina x neyrautii* Foucaud (Baumel et al., 2002a). The base chromosome number is x=10 in the genus *Spartina*, 2n ranging from 40 to 124 for species investigated to date (Marchant, 1970). Both polyploidy and aneuploidy are observed in this genus (Fig 1).

Mobberley (1956) and Partridge (1987) reported that *Spartina* geographically distributed along the east coast of North and South America, with small populations found on the west coast of North America, Africa and Europe. *Spartina* species are salt tolerant perennial rhizomatous wetland grasses that can be found primarily in wetlands, especially estuaries, but will also establish in internal freshwater habitats (Mobberley, 1956; Partridge 1987).

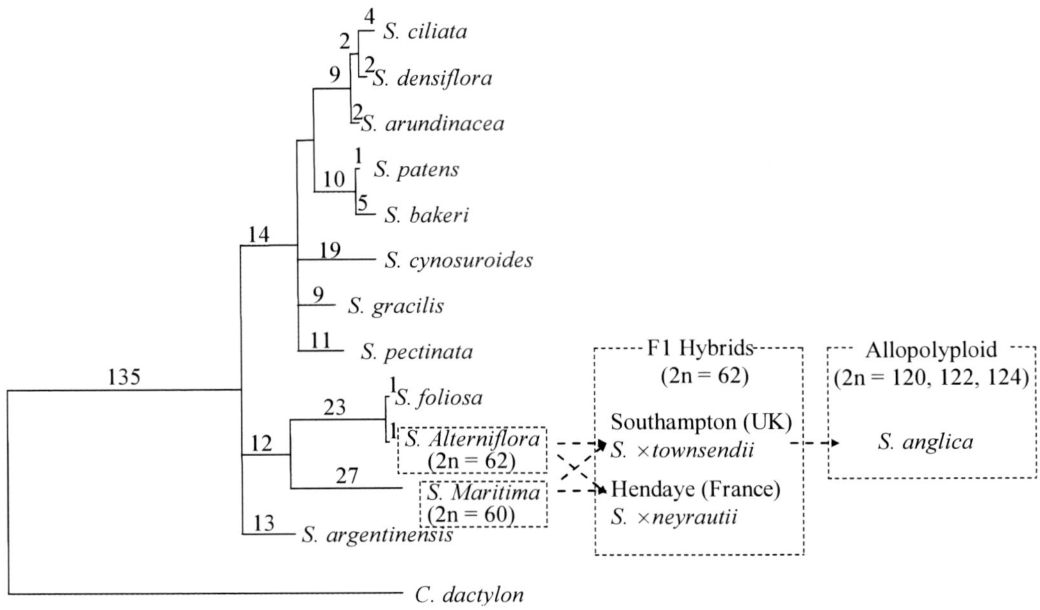

Figure 1. Phylogeny of Genus *Spartina* Schreb. (Poaceae). (Wang & An, 2006)

2.3 Cytogenetics of *Spartina alterniflora*

The chromosome numbers of *S. alterniflora* existed a high degree of polyploidy with chromosomes present in multiples of seven (Church, 1940). Octoploids with 56 and the decaploids with 70 chromosomes have been reported, and two levels of polyploidy within S. alterniflora were found to be morphologically and ecologically distinct. Thus, Church (1940) concluded that the basic number of chromosomes within this genus is seven. However, controversy existed as to whether the tall and short forms of *S. alterniflora* differ genetically or simply reflect differences in adaptation to the environment to which they are exposed. The short form, *S. alterniflora* var. glabra (Muhl.), was characterized as being octoploid (n=7) with 56 chromosomes and the tall form, *S. alterniflora*, var. pilosa (Merr.), was characterized by being a more robust decaploid (n=7) with 70 chromosomes. A cytological study of *S. alterniflora* populations in northeastern U.S.A. and Canada conducted by Marchant (1970), excluded *S. alterniflora* from the basic number of 10 and affirmed that both the short and tall forms have the same chromosome number of 62. This is unusual with respect to the basic number of x=10 generally found in the genus. It is not known where the two additional chromosomes in the complement of *S. alterniflora* originated. Compared with the basic number observed in other species, it is possible that they represent a tetrasomic condition.

2.4 Reproductive Biology of *Spartina Alterniflora*

Flowers of *S. alterniflora* are smooth, lack hairs, and panicles are erect to arching. The panicles are usually 10 to 45 cm long with 5 to 30 spikes alternately arranged and appraised to main axis with 10 to 50 sessile spikelets along one side of the axis of each spike. Under favorable conditions, *S. alterniflora* can reach sexual maturity in three to four months (Smart & Barko, 1982). *S. alterniflora* responds to short day length for plant growth and flowering (Seneca and Broome,1972). Flowering occurs earlier under short-day than under long-day conditions in 22-26℃ and 26-30℃ temperature conditions. Also, Seneca (1974) found considerable morphological and physiological variation among populations of *S. alterniflora* from the Atlantic and Gulf coasts. Whereas southern populations flowered later and exhibited a longer growing period, they showed less sensitivity to changes in photoperiod than northern populations.

S. alterniflora flowers annually with variable flowering dates throughout its geographic distribution (Mobberley, 1956), and the species is protogynous, meaning that female flowers mature before male flowers, and this strategy helps ensure outcrossing. Stigma protruded from the floret 2-5 d before anther and pollen germinated about 15 min after contacting the stigma, and pollen tubes grew to the micropyle within 55 to 75 min (Fang et al., 2004). Daehler and Strong (1994) reported that inflorescences of self-pollinated plants generally had lower seed set than that of controls, and none of the self-pollinated seed were viable. Spikelet viability was not a function of clone size but there was some genetic control of spikelet viability shown by the uniformity of seed set that existed within large clones, and there was inbreeding depression with different degree among clones, which may be due to the *S. alterniflora* populations on the west coast were established from a relatively small number of genetic individuals (Daehler and Strong, 1994).

According to Mooring et al. (1971), seeds are set in fall and subsequently dispersed but remain dormant until early the next spring. The phenomenon of dormancy can help seed survive through a cold winter. Seed output of both native and introduced *Spartina* populations has been found to be highly variable (Callway and Josselyn, 1992). Seed production in *S. alterniflora* marshes has also been found to be unpredictable across years (Broome et al., 1974), and the plant does not produce seed in several areas where it has been introduced. No flowers have been observed in New Zealand or in Padilla Bay (Pacific Northwest coast of the U.S) populations for almost 50 years after its introduction (Partridge, 1987). Low soil temperature can delay or suppress flowering and reduce seed production in *Spartina*. Callaway and Josselyn (1992) reported that seed production was very erratic as evidenced by the variation in the number of seeds produced per flower, ranging from 3% to 20% during a two year period of evaluation. The germination rates ranged from 0% to 59% among clones (Daehler and Strong, 1994). Variable germination rates were also observed by Callaway and Josselyn (1992) with roughly 37% of seeds collected in San Francisco Bay. Sayce (1988) found only a 0.04% germination rate for seeds collected in Willapa Bay. However, Qin et al. (1985) reported that the germination rate of seeds of Florida form (F-form), which introduced from Florida state, USA, was over 90% with distilled water, and that of other two ecotypes, Georgia form (G-form) and North Carolina form (N-form) which introduced from Georgia state and North Carolina state respectively, were 76% and 66%. In addition, there were significant differences among the germination potential of three ecotypes. The results indicated that the germination characteristics of *S. alterniflora* seeds varied with the sources of seeds.

2.5 Distribution of *Spartina Alterniflora*

Smooth cordgrass (*S. alterniflora*) dominates the lower intertidal zone throughout most of its native range, extending across the entire northern Gulf of Mexico and up the Atlantic seaboard of North America as far as Quebec and Newfoundland. Due to its role as an ecosystem engineer, *S. alterniflora* represents an important economic resource, both for the critical habitat it provides to numerous avian and fisheries species, and for the sediment stabilization provided by its dense rhizomatous networks developed through clonal propagation. *S. alterniflora* had been deliberately introduced into many country for the control of coastal erosion and land claim (Chung et al., 2004). Thus it distribution consequently enlarged to include Asia, Europe and Oceania.

In its native range, *S. alterniflora* typically dominates high salt marshes, growing from 0.7 m below mean sea level to approximately mean high water (Bertness, 1991). In Willapa Bay, located on the Pacific Northwest coast of the U.S., the plants have been observed growing between 1.8 and 2.8 m above mean lower low water (MLLW), and transplants have been known to survive within 1.0 m above MLLW (Sayce, 1988). The species are also found in inland freshwater habitats.

3. The Ecological Superiority for Successful Invasion

The ability of this species to trap large volumes of tidal sediments led to its deliberate introduction in several parts of the world (North Europe, Australia, New Zealand, China…) in conjunction with land reclamation. However, the rapid expansion of the invasive species is now considered to be a threat to coastal environments (Daehler & Strong, 1996; Ayres et al., 2004; Chen et al., 2004). *S. alterniflora* have become model organisms for studying biological invasions from both ecological and genetic perspectives (Blum et al., 2004). The invasive mechanism of the species became involuntarily a researching hotspot.

3.1 The invasive patterns of *Spartina Alterniflora*

All of invasive species are introduced by chance or deliberately into a region or country by human being. However, the fates of invasive species vary with different invasive pattern and passages. In the last two hundred years, *S. alterniflora* was introduced many countries in the world, and its distribution has expanded from Atlantic coast of America to Europe, western coast of North America, New Zealand and coast of China (Baumel et al., 2001).

In England, *S. alterniflora* was introduced by chance in the Southampton area from east coast of North America in 1816, and then the species dispersed to British and Ireland where it hybridized with the native Spartina maritime Fernald, resulting in a sterile perennial hybrid named *Spartina townsendii* H. & J. Groves. *Spartina townsendii* still remains localized in the locality of Hythe (Raybould et al. 2000). Chromosome doubling in the hybrid gave rise to a new fertile and robust rhizomatous allopolyploid species, *Spartina anglica* C.E. Hubbard, which has been actively colonizing British and French salt marshes and estuaries since its formation around 1890 (Baumel et al. 2001). In France, *S. alterniflora* was also introduced during the 19th century in Brittany near Brest (Elorn Estuary), and in the Basque region near Hendaye (Bidassoa Estuary). In Hendaye, *S. alterniflora* hybridized with native Spartina species, S. maritime, and produced hybrid *S.* × *neyrautii*. In England and France, the *S. anglica* was rapidly expanding, and the distribution of *S. alterniflora* was significantly reduced. The species only remained in local sites, such as the sole remaining site of *S. alterniflora* in Britain was at Marchwood (Raybould et al., 1991).

S. alterniflora plant parts or seeds probably contaminated barrels used to pack oyster spat and young adults for shipment to Willapa Bay, Washington in the 1800's and early 1900's. 9% of the total intertidal habitat in Willapa Bay was infested by the species. Experimental studies for control of this weed by State of Washington and federal agencies were in the late 1980s. The expansion of *S. alterniflora* populations introduced accidently was at a lower level and local distribution for Allee effect (Davis et al., 2004a, 2004b; Taylor et al., 2004).

The invasion of *S. alterniflora* resulted from purposive introduction rather than involuntary infection. A population of *S. alterniflora* had been identified in the late 1930s, and the Oregon Department of Agriculture began control efforts in 1990. The infestation was finally eradicated in 1997 by digging. *S. alterniflora* was deliberately planted in San Francisco Bay in 1973 as part of a marsh reclamation project near the town of Fremont. Five years later, almost ten thousand plants of *S. alterniflora* were transplanted from Fremont into the shoreline of Alameda Island, and later to San Bruno. During the 1980s, restoration projects throughout the Bay may have further spread the species. Seeds and individuals of three ecotypes of *S. alterniflora* were introduced from North Carolina, Florida and Georgian on the east coast of America to China respectively. Initial plant materials were planted in Luoyuan County of Fujian Province, China. In the after years, the seeds or / and vegetative individuals were transplanted from Luoyuan County to coastal provinces of China. The area of the *S. alterniflora* was explosively expanding from 2.6 km^2 in 1985 to 1,120 km^2 in 2002. Thus, *S. alterniflora* as one of sixteen species was placed on the first invasive species list by State Environmental Protection Administration of China (SEPAC) (Wang et al., 2006).

From the above discussion, one may concluded that purposely planting is the sufficient condition of *S. alterniflora* invasion. In China, after many times and distributed introduced, the expansion of *S. alterniflora* population exhibits a multi-point outbreak pattern.

3.2 Genetic Basic and Ability of Hybridization and Introgression of *Spartina Alterniflora*

There are four broadly distributed populations along the Atlantic and Gulf coasts of the United States (New England, the Mid-Atlantic, the South Atlantic and the Gulf coasts). After assess the genetic variability in tall form *S. alterniflora* collected the two regions using randomly amplified polymorphic DNA (RAPD), O'Brien et al. (1999) classified those populations into three geographic areas (New England, South Atlantic, and Gulf coast), closely approximating the three types defined by Seneca (1974).

The results of a comparative study on genetic variation of population collected from Gulf Coast (introduced populations)) and Eastern Coast of the USA (native populations) respectively indicated that *S. alterniflora* populations had a high genetic diversity, and intrapopulation genetic diversity was higher than that of interpopulations collected from different region (Perkins et al., 2002). There was a positive relationship between genetic variation and spatial distance of populations, and the spatial distribution of genetic structure was a continuum rather than a discrete regime (O'Brien et al., 1999). Furthermore, moderate disturbance could increase genetic diversity of populations (Edwards et al., 2005). For introduced populations with a long time history, its genetic diversity exhibited a lower level (Blum et al., 2004); however, the genetic differentiation maybe happened during the early stage of invasion (Daehler, 1998). It is an exemplum that a dwarf ecotype has evolved and spread since the introduction of *S. alterniflora* to San Francisco Bay from Maryland in the 1970s (Daehler et al., 1999). *S. alterniflora* as a notorious invasive species, the genetic diversity of the species was lower in variously spatial scales, whereas the genetic differentiation of intrapopulations was 22% (Deng et al., 2007). Thus, genetic differentiation have happened in the process of population expansion, and the ability of rapidly adaptive evolution resulted from the differentiation may be one of the reasons that the species widely expanded in a high rate.

Hybridization and introgression also are the basics of outbreak of invasive species. Hybridization between species or between disparate source populations may serve as a stimulus for the evolution of invasiveness (Ellstrand & Schierenbeck, 2000). After introduced into England, *S. alterniflora* hybridized with native *Spartina maritima* and resulted in a sterile hybrid *S.* × *townsendii*. Chromosome doubling in this hybrid gave rise to a new fertile allopolyploid species, *Spartina anglica* (2n = 122–124), a vigorous and aggressive perennial plant that has been actively colonising British salt marshes since its formation (Raybould et al., 1991). *S. anglica* displays wider ecological amplitude than its parents across the successional sequence of salt marsh zones (Thompson, 1991) and rapidly spread along the West-European coast(Baumel et al., 2001). Contrast to native *S. foliosa, S. alterniflora* had greater mall fitness, and the hybrids appeared vigorous and were recruiting more rapidly than their parents (Daehler & Strong, 1997; Anttila et al., 1998). Furthermore, hybrids could colonize the habitats where their parents were absent (Ayres et al., 1999). Introgression led to genetic pollution of native *Spartina* species and enhanced the invader (Ayres et al., 2008).

3.3 Ecological Adaptability of *Spartina Alterniflora*

S. alterniflora has strong adaptability and tolerance on stress environment with broad distribution (30°-50° N). Furthermore, the species can grow well in different substrate (Xu & Zhuo, 1985) and have a large degree of plasticity and create some growth forms in different habitats. These growth forms have a genetic basis, either in the basic coding information or regulatory mechanisms that becomes permanently set at an early development stage (Gallagher et al., 1988). However, 55% phenotypic variability of *S. alterniflora* is due to environmental variation (Richard et al., 2005). Salt marsh is one of the most stressful habitats in the world. There are direct relationship between successful expansion and outbreak of *S. alterniflora* and its powerful tolerance to the environmental factors of this stressful habitat, and the powerful tolerance consequently resulted to its competitive superiority (Pennings et al., 2005).

3.3.1 Salinity Tolerance

S. alterniflora is a facultative halophyte. It can tolerate a broad range of salinities from near fresh water to ocean concentrations. Salt crystals can often be seen on the leaves during the growing season. The highest growth rate occurred at salinities of 20‰ or less, with the upper limit for salt tolerance being 60‰. *S. alterniflora* could adopt multi-approach to tolerate salinity stress. In salinity habitats, the species tolerate the most of salinity stress by exclusion, then excretion and accumulation (Bradley and Morris, 1991). In addition, *S. alterniflora* has also been shown to combat high intracellular salinity with osmotic adjustment (Cavalier, 1983). Furthermore, the cell structure of the species presents high stability to salinity stress. Only part of mitochondria structure of *S. alterniflora* seedling had been damaged under 200 m mol/L NaCL treatment, and the treatment did not effect the other cell organelles and cell wall ultramicrostructure.

3.3.2 High tolerance to Flood and Hypoxia Stress

The salt marshs are flooding regularly by tide. Flooding results in plant hypoxia and leads to the reduction of photosynthesis, and consequently decreases plant growth. The extensive aerenchyma system of *S. alterniflora* may supply submerged portions of the plant with atmospheric oxygen as well as lower metabolic demands of the plant (Maricle & Lee, 2002). Furthermore, the efficiency of oxygen transport exhibits Matthew Effect. Dense growth and expansion facilitate the tolerance of hypoxia and enhance the invasiveness of the species (Bertness, 1991). The species can stand 12 h inundation each tide (Xu & Zhuo, 1985). The root growth was restrained when oxidation-reduction potential (ORP) of soil was lower than +350 mv, and it was seriously impacted when ORP under +200 mv (Pezeshki, 1997).

3.3.3 Competitive Superiority

As marsh elevation increases to landward, tide related abiotic factors, such as the frequency and duration of inundation, are reduced and competition for resources becomes increasingly important in structuring plant communities (Gray, 1992; Huckle et al., 2002; Daehler, 2003). In inundated and high salinity habitats, *S. alterniflora* showed significantly competitive superiority (Wijte & Gallagher, 1996a, 1996b). The level of nitrogen nutrition affects the result of competition of *S. alterniflora* with co-existed species. Lower nitrogen level was disadvantageous for *S. alterniflora* growth, whereas, the environmental pollution and eutrophication increased the nitrogen nutrition and enhance the species invasion and expansion (Levine et al., 1998; Tyler et al., 2007; Zhao et al., 2007).

High sulfidic concentration in salt marsh limited the growth of some plant species. Increased sulfide in the rhizosphere reduces the ability of *Phragmites* to take up nutrients. However, *S. alterniflora* are better-adapted to sulfidic soil conditions (Stribling, 1997; Chambers et al., 1998). The presence and distribution of sulphate Bacteria on the roots and rhizomes of *S. alterniflora* were more than that of other salt marsh species, and high sulphate reduction rates (SRR) could mitigate negative action of sulfide (Nielsen et al., 2001). Furthermore, *S. alterniflora* could exclude the competitor through releasing sulfide to soil. The sulfide concentration in sediment of *S. alterniflora* communities was more than ten times of that of *Phragmites australis* communities (Chambers et al., 1998; Seliskar et al., 2004).

3.3.4 Constructing Mutualism

The relationships between plants always vary with habitats change, i.e. the relative importance of competition and mutualism varies with environmental condition (Brook & Callaghan, 1998; Callaway & Aschehoug, 2000). The relationship between plants presents competition rather than mutualism in better habitats, and the reversed relationship is more common in stress conditions (Bertness & Callaway, 1994; Bruno, 2000; Callaway & Penning, 2000). *S. alterniflora* could benefit through the mutualism relationship when it coexists with *Ascophyllum nodosum*. The latter provides more nutritions for *S. alterniflora* to increase the biomass accumulation, reversion, and the shade of the canopy of *S. alterniflora* decrease the leaf-surface evaporation of *Ascophyllum nodosum* (Gerard, 1999).

The symbiosis of *S. alterniflora* and fixed nitrogen bacteria could mitigate the limitation of poor nitrogen soil condition (Dai & Wiegert, 1997; Boyer & Zedler, 1998; Piehler et al., 1998; Tyler et al., 2003). The symbiosis between *S. alterniflora* and sulfidic bacteria could improve the rhizosphere microhabitat (Nielsen et al., 2001).

3.4 Reproductive Capability of *Spartina Alterniflora*

Seed dispersal is one of modes with which *Spartina alterniflora* population colonized new habitats (Richards et al., 2004; Davis, 2005). The maximal number of grain per inflorescence was 665 (Fang et al., 2004). After shedding at autumn, seeds of *S. alterniflora* keep dormancy to spring of second year and the maximal germination rate is over 90% (Qin et al., 1985). Dormancy mechanism will be favor of seed survival through the severe winter, and companying with temperature rise and rainfall increase, seeds are germinating. The strategy could increase the survival of seedling. In hypoxia condition, the strategy which coleoptile and hypocotyl elongate quickly will accelerates seedling of *S. alterniflora* coming up out of land. Moreover, oxygen produced by the photosynthesis of green embryo in coleoptile could stimulate the root growth of the species (Wijte & Gallagher, 1996a, 1996b).

S. alterniflora could expand the populations using rhizome or vegetative fragment dispersal besides sexual reproduction (Daehler & Strong, 1994). The population maintenance and expansion was depending fully vegetative reproduction in New Zealand (Partridge, 1987). In foreland of invasive populations, the isolated "pioneer" clones joint into continuous swards through vegetative reproduction (Davis et al., 2004a). However, the vegetative reproduction and clonal expanding rate vary with habitats. The length of rhizome and capability of vegetative reproduction varied with the organic content in substrate, and 14 new shoots were produced by each seedling in higher organic content substrate (Padgett & Brown, 1999). In large bare patches in field, *S. alterniflora* seedlings grew non-directionally, and produced as many as 36 tillers in one growing season (Metcalfe et al., 1986). Lateral growth rates of individual *S. alterniflora* patches increased linearly at 79.3 cm cm yr^{-1} and recruitment of new clonal patches had been episodic and increasing in frequency (Feist & Simenstad, 2000). Proffitt et al. (2003) reported that the rate of clonal expansion in diameter was 3.1 m per year. The number of new shoots produced by a ramet, which planted in Luoyuan County, Fujian Province of China, was more than one thousand in one year. The rate of clonal expansion in diameter was more than 2 m and the diameter of clone was more than 19 m after 4 years (Chung et al., 2004). In a word, *S. alterniflora* not only produces a great deal of seeds and colonizes new habitats, but also expands its population through rapidly clonal reproduction. *S. alterniflora* population practise local movement ("flow" diffusion) with clonal reproduction and long-distance jump with seeds (Hastings et al., 2005).

3.5 The Invasibility of Native Ecosystem as a Necessary Condition of *Spartina Alterniflora* Invasion

Before the invasion of *Spartina*, there are few plant species distributed in the tidal zone for the severely physical condition. Thus, *S. alterniflora*, as a pioneer species of community succession, occupies an empty niche at the early stage of invasion (Chung et al., 2004). After successful establishment, the populations have a chance to rapid expansion. Such as in China, all of 90% foreshore of Jiangsu Province is silty, and the total area is 6,533 km2. Previous of *Spartina* introduced, plants only distributed in the zone above mean higher water level (MHWL), and the broad foreshore between MHWL and mean water level (MWL) and under MWL where *Spartina* could colonize and establish population was empty (Guan et al., 2003).

Contrasts to *S. alterniflora*, the native species are inferior competitors. The growing dominance of *S. alterniflora* restrained the growth of *Scirpus mariqueter* when the two species were competing. Consequently, *S. alterniflora* successfully invaded into *S. marigueter* community (Chen et al., 2004). The photosynthesis rate of *Phragmites australis* was significantly lower than that of *S. alterniflora*, and this disadvantage led to its loss when P. australis competed with *S. alterniflora* (Zhao et al., 2005; Zhao et al., 2008).

Spartina alterniflora, as a successful invader, can enhance the ecosystem carbon and nitrogen stocks and therefore facilitate its invasion. The species invasion altered ecophysiological processes, resulted in changes in NPP and litter decomposition, and ultimately led to enhanced ecosystem C and N stocks in invaded ecosystems (Liao et al., 2007, 2008). Furthermore, *S. alterniflora* accelerate sedimentation through its extensive rhizome and upright stems, at the same time, burial by sediment stimulate its growth and reproduction (Donnelly & Bertness, 2001; Morris et al., 2002).There is a positive feedback relationship between sedimentation and growth of *S. alterniflora* in coastal salt marshes occurs. A self-facilitation mechanism may be developed during the process of *S. alterniflora* invasion (Deng et al., 2008).

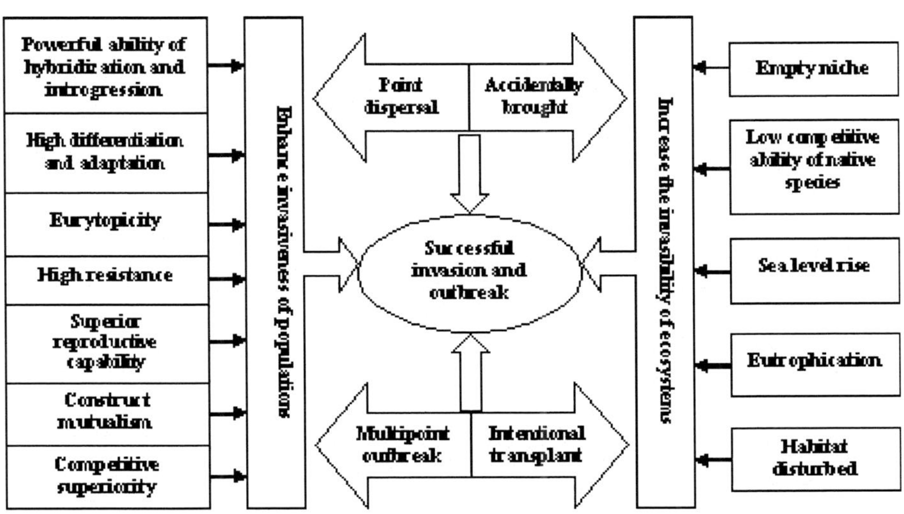

Figure 2. The invasive mechanism of *Spartina alterniflora* (Deng et al., 2006)

In a word, the invasive mechanism of *S. alterniflora* involves three factors: the invasiveness of invasive species, the invisibility of ecosystem and offering an opportune invasive passage (Fig 2). The invasiveness of *S. alterniflora* includes powerful ability of hybridization and introgression, mighty adaptability and superior reproductive ability, eurytopicity and competitive superiority. The ecosystems with empty niche, inferior native competitors and frequent disturbance, such as foreshore, are the necessary conditions of *S. alterniflora* invasion. The accident dispersal and purposed transplant of propagules of the species directly result in the invasion. Especially, in China, large-scale transplants accelerate the invasive rate and enlarge the explosive scope of *S. alterniflora*. The outbreak of the species represents a characteristic with point-resource dispersal and multi-loci expansion.

4. The Functional Evaluation and Management of *Spartian Alterniflora*

Due to its valuable ecological and economic functions, *S. alterniflora* had been worldwide introduced (Chung, 2006). However, the species also resulted in a series of negative effects on ecosystem health and economic development of introduced regions or countries for its robust reproductive capability and rapid expansion (Daehler & Strong, 1996; Hedge et al., 2003). The argument about function of S. alterniflora is going on for a long time (Qin et al., 1997; Hedge et al., 2003; Sun et al., 2004; Chung et al., 2004, 2006).

4.1 The Functional Evaluation on *S. Alterniflora*

4.1.1 Beneficial

In its native habitat, *S. alterniflora* is valued. The species is highly productive, exporting approximately 1300 $g{\cdot}m^{-2}$ of detritus annually to the estuarine system (Edwards & Mills,2005). Moreover, *S. alterniflora* is highly regarded for erosion control, as well as fish and wildlife values in its native range (Simenstad & Thom, 1995). In the east coast of America, some waterfowl and wetland mammals eat the roots and shoots of this plant. In addition, stands of *S. alterniflora* can serve as a nursery area for mangroves, and estuarine fishes and shellfishes.

There are also some economically beneficial uses for *S. alterniflora*. The species is palatable to livestock, so the plant's continued spread may increase available pasture. Efforts have also been made to use *S. alterniflora* in paper production (Qin & Chung, 1992; Qin et al., 1997).

As an "ecological engineer", *S. alterniflora* colonizes on the severe habitat as pioneer species of community succession and transforms open mudflats into densely vegetated marshes (Travis et al., 2002, 2004, 2005). Consequently, coast is stabilized and erosion is controlled. In addition, *S. alterniflora* plantation could be a countermeasure against disasters of sea level rise with the siltation and salt marsh elevation, as well as the methods to produce lands for the tidelands cold been reclaimed. Furthermore, the function of *S. alterniflora* have been explored abundantly in China, such as soil amelioration, pollution control, animal fodder, green manure, fish feed, primary producer of detritus food webs, food additive, etc. (Chung, 2006).

4.1.2 Negative Impacts

Firstly, invasion and expansion of *S. alterniflora* results a series of negative ecological impacts. *S. alternifolra* may displace native plants, such as *Zostera marina* (seagrass) at lower elevations, which provide important refuge and food sources for fish, crabs, waterfowl, and other marine life (McMillan et al., 1995), and *Salicornia virginica*, *Triglochin maritimum*, *Jaumea carnosa*, and *Fucus distichus* at higher elevations (Simenstad and Thom, 1995). As *S. alterniflora* population expanding, shorebirds and waterfowl will lose potentially important foraging and refuge habitat. In the Willapa National Wildlife Refuge, *S. alterniflora* has displaced an estimated 16 to 20 percent of critical habitat for wintering and breeding aquatic birds (Foss, 1992). At the same time, bottom-dwelling invertebrate communities of mudflats will be replaced by salt marsh species. Benthic fauna of *S. alterniflora* meadow is generally depleted in comparison to nearby mudflats, and its invertebrate opulations in the sediments are smaller than populations in intertidal mudflats (He et al., 2007). Juvenile chum and Chinook salmon may lose prey resources and other important attributes of mudflat nurseries (Simenstad & Thom, 1995). In short, mud- and sandflat communities based on bottom-dwelling microalgae will decline, being replaced by food webs driven by the supply of *Spartina* detritus (Simenstad and Thom, 1995).

Secondly, some of the very traits that make *S. alterniflora* valued are also the greatest causes for concern. *S. alterniflora* dense and tall stems and leaves slow tidal water, thus trapping sediment. As a result, *S. alterniflora* causes tidelands to raise more than they would if they were unvegetated or vegetated by other species (Thompson 1991; Chung et al., 2004). Sedimentation rate was up to 260 mm/year in China (Chung et al., 2004). The sedimentation with such rate alters the topographic and hydrographic characteristics of coast, i.e. before *Spartina* colonization, foreshores or estuaries consisted primarily of bare, gently-sloping mudflats with shallow tidal channels, *Spartina* replaces gradually sloping mudflats with badly drained marshes that commonly have steeply sloping seaward edges and deep, steep-sided channels (Gray et al., 1991). In addition, infestations of *Spartina* may block some navigational channels. The large and dense populations at or in river mouths may cause particular problems by decreasing flow and leading to increased flooding, especially during periods of heavy precipitation and/or above normal tides.

Thirdly, the expansion of *S. alterniflora* also produces some detrimentally economical effects. The habitats of many native animal species, including some endangered birds and economic mollusks, such as *Bullacta exarata*, were threatened by *S. alterniflora* invasions (Luiting et al., 1997). In addition to wildlife impacts, *S. alterniflora* poses threats to a multi-million dollar shellfish industry in a local economy (Grevstad et al., 2003).

Lastly, changes associated with *S. alterniflora* also impact recreation. Loss of beach habitat and navigation routes, reduced water access, and other alterations to the estuarine ecosystem may result from the spread of *S. alterniflora*. Therefore, activities, such as fishing, hunting, boating, bird watching, botanizing, and shellfish harvesting, that are dependent on the extant intertidal ecosystem could be negatively impacted by the continued spread of *Spartina* (Ebasco Environmental, 1993).

In general, the functions of *S. alterniflora* should be objectively evaluated according to the local hydrographic and geographic characteristics. Just as to understand the different east/west perspectives on the values of *Spartina*, it is important to recognize that Atlantic/Gulf coast estuaries are fundamentally different than their Pacific Coast counterparts. The Pacific Coast is macrotidal, while the Atlantic/Gulf coasts are microtidal (Simenstad & Thom, 1995). In addition, the Pacific Coast is more geologically active, and tectonic activity has a much greater influence on coastal marsh systems. Subsidence due to compaction of marsh soils that results from insufficient sediment supply is less important on the Pacific Coast (Thom, 1992). Finally, on the East Coast, the prevailing wisdom is that salt marshes are the key to productivity of estuarine systems because of the contribution of their detritus. However, on the West Coast, secondary productivity from tidal mudflats is more important than detritus exported from salt marshes. The sustainable development of China is severely limited by the inconsistency of supply-demand for land, and the reclamation of tideland can effectively mitigate this condition. The main functions of *S. alterniflora* are coastal stabilization, erosion control and reclamation. Thus, beneficial function of *S. alterniflora* is larger than its negative impacts from the view of national macroscopical economics. On the standpoint of ecology and local economic development, the invasion of *S. alterniflora* is a nuisance rather than a friend.

4.2 Management of *S. Alterniflora*

The experimental researches and practices on control and management of *S. alterniflora* are carrying through, although the functional evaluation of the invasion of the species could come to an agreement (Grevstad et al., 2003; Hedge et al., 2003; Major et al., 2003). The main eradiation or control methods include mechanical method, chemical method, biocontrol method and comprehensive eco-control method.

4.2.1 Mechanical Methods

The mechanical methods is comprised of hand-pulling or manual excavation, covering with special cloth, creating dike to flooding or draining, continuing mowing, flaming, etc.. The effect of hand-pulling is significant for seedling of *S. alterniflora*. While, for clones or swards of *S. alterniflora*, it is difficut to pull or excavate by hands. Furthermore, the clones distribute on mudflats and the substrate is soft. Thus, the feasibility of hand-pulling or manual excavation is very small. Experimental results indicated that covering with special cloth for succession two growing season could effectively control small patch of *S. alterniflora* (*Spartina* Task Force, 1994). Continuing mowing could restrain the growth. However, it is necessary to mow nine to ten times from spring to autumn, and it must be three to four years for fully eradication of *S. alterniflora*, and the strategies directly influence the control effects (Taylor C. M. & Hastings, 2004).

4.2.2. Chemical Methods

The chemically methods is using herbicide to control *S. alterniflora*. In the chemical control practice in an estuary in the USA, N-(phosphonomethyl) glycene (glyphosate) was the most herbicide registered for aquatic use. The second was the solution of Arsenal (±-2-[4,5-dihydro-4-methyl-4-(1-methylethyl)-5-oxo-1H- imidazol-2-yl]-3-pyridinecarboxylic acid) (Patten, 2002, 2003). The effects of herbicides varied with time and place when or where herbicides were used. Moreover, the effectiveness of herbicides on *Spartina* infestations appears to be highly variable with considerable variation between application methods. Low volume aerial application at permitted concentrations, although relatively inexpensive at approximately $420 per ha, has been far from impressive with combined mean efficacies of approximately 30% (Major et al.,2003). Wicking and wiping methods have also produced variable results. These methods are not cost effective and only suitable for small infestations. High volume hand held spray applications have been the preferred Rodeo application methods for *Spartina* control in Washington. Some trials report that Rodeo produces efficacies from zero to 50% (Patten 2000). However, others (Crockett 1997) report efficacies ranging from 85 to 97% control with 1 to 5% solutions of Rodeo sprayed to wet. The major parameters influencing efficacy appear to be the interaction of tidal elevation and period of time from post-spraying to tidal inundation and leaf cleanliness (Crockett, 1997; Patten, 2003).

Herbicides are noxious, and they maybe directly or indirectly impact to the health of human being, fauna and flora (Chen et al., 2004). Thus, the further studies are recommended for herbicide application for *S. alterniflora* controls.

4.2.3 Biological Control

The looking for and releasing natural enemy become one of the hotspots about biological invasion due to the advantages of biological control, such as durative efficacity, inexpensive cost, and ecological safety. The most promising biocontrol agent appears to be a Homopteran plant hopper (*Prokelesia marginata*) that feeds on the vascular fluids of *Spartina* species by piercing the leaf with its stylet (Daehler & Strong, 1995). The other animals in biocontrol practice are *P. dolus, Delphacodes penedetecta* (Hemiptera: Delphacidae) (Huberty & Denno2001; Ferrenberg & Denno, 2003). The ergot fungus, Claviceps purpurea, also has potential as a biocontrol agent. *Littoraria irrorata*, *Melampus bidentatus*, *Geukensia demissa, Orchelimum fidicinum*, *Conocephalus spartinae* (Silliman & Zieman, 2001; Grevstad et al., 2003; Pennings & Silliman, 2005). However, some questions are uncertain for the anfractuous interactions between / among biology, such as whether natural enemy will impact to other biology besides *Spartina*, and introducing natural enemy probably results to new biological invasion. There are arguments about natural enemy introduction, and the status of biocontrol for *Spartina* is experimental researches.

4.2.4 Ecological Control

The principle of ecological control is that according the regular of community succession, using native species with economic or ecological value to replace invasive species through altering the habitat conditions. Bonilla-Warford & Zedler (2002) reported that invasive species *Phalaris arundinacea* had been replaced successfully by native species *Spartina pectinata.* The growth of *Spartina* had been restrained by transplanting *Sonneratia apetala* in Aoqidao of Zhuhai, Guangdong Province, China (Wang et al., 2003). There are some experimental practices to control *S. alterniflora* in the salt marsh of Jiangsu Province, China through constructing dike to restrain the dispersal of rhizome of *S. alterniflora*. At the same time, irrigation with fresh water reduces soil salinity and reconstructs the *Phragmites australis* wetlands.

4.2.5 Integrated Control

The ecological and economic consequences of *Spartina* invasion varies with the local geographical characteristics and the process of oceanic dynamics. According the results of San Francisco Bay and Willapa Bay, it is very difficult to fully eradicate *S. alternilora* (Daehler & Strong; 1996). Thus, a more ecosystem-based approach to management, including control, needs to recognize the overall, long-term ecological role of *S. alterniflora* in the world (Hedge et al., 2003; Majoret al., 2003, 2004). The strategies include limiting the distribution of *S. alterniflora* through ecological engineering, and at the same time, exploring the economical value of the species. Total flavonoids and biologically mineral liquor could be distilled from the stems of *S. alternilora* (Qin & Chung, 1992), and the residue could be used as culture medium of mushroom. The strategies not only avoid pollutions but also increase the output of the ecosystem (An & Chung, 1991, 1992; Qin & An, 1998).

5. Conclusion

S. alterniflora is a vigorous and invasiove species. Its vigorous growth form allows it to exploit a specific habitat niche where is a large areas of unvegetated mudflats. At the same time, the robust stems and reticular rhizomes and roots of *S. alterniflora* makes it suitable for shoreline stabilisation, land claim and erosion defence purposes. Thus the species has been artifically spread from its native habitat to a large number of estuarine ecosystems throughout world. Unexpectedly, the quick expansion and outbreak of the species resulted some negatively economic and ecological effects in the last twenty years. *S. alternilfora* invasion mechanism and control strategies have been a hot topic.

The previous researches showed that powerful ability of hybridization and introgression has been a genetic basic, superior reproductive capability has been the sources and strong ability of anti-stress and adaptability has been an ecological and physiologic basic for *S. alterniflora* invasion and expansion, respectively. In most invasive habitats, the expansion of *S. alterniflora,* based on intentionally transplants, indicated the mode of point dispersal. The episodic and continuous dispersal pattern of seeds has been playing an important role for maintain, recruitment and outbreak of *S. alterniflora* population. Meanwhile, consecutive expansion of *S. alterniflora* populations was ensured by the trait of potently clonal growth. Meanwhile, a large number of researches about *Spartina* management and control method have been carried out.

In general, the species of *S. alterniflora* is a doouble-edged sword. Its growth and expansion can bring some economical benefit and positively ecological effects. However, the immoderate invasion of the species also can lead to a series of negative issues on economic and ecology. While the exact ecological and economic consequences of these changes are uncertain, the potential for damage may be extensive. The only way to test the full extent of the impacts would to wait until *S. alterniflora* is widely established, at which point, the species would be virtually impossible to control. Therefore, prevention of seed production in all designated areas is required to help contain this species and prevent its further spread. At the same time, the integrated strategies with exploring the economical value and other restraining growth of *S. alterniflora* methods should be adopt to manage the biological invasion.

ACKNOWLEDGMENTS

Founded by NSFC (No.30400054), Natural Scientific Foundation of Jiangsu Province, China (BK2007152) and Natural Scientific Foundation of Nantong University (07Z031) and Research Foundation for Doctors of Nantong University (08B12).

REFERENCES

An S. Q. & H. C. Chung (1991). Ecological engineering with *Spartina spp.* plantation and benefit analyses of prevention of disasters. *Journal of Nanjing University Special Issue*: 323-328.

An S Q & Chung C H. (1992). Types, distribution and utilization of major seashore ecosystems in China. In: Qin P and Chung C H eds. *Applied studies on* Spartina. Beijing: Ocean Press. 21~27.

An S. Q., Gu B. H., Zhou C. F., et al. (2007). *Spartina* invasion in China: implications for invasive species management and future research. Weed Research, 47: 183-191.

Anttila C. K., Daehler C. C., Rank N. E. & D. R. Strong (1998). Greater male fitness of a rare invader (*Spartina alterniflora*, Poaceae) threatens a common native (*S. foliosa*) with hybridization. *American Journal of Botany*, 85(11): 1597-1601.

Ayres D. R., D. G. Rossi, H. G. Davis & D. R. Strong (1999). Extent and degree of hybridization between exotic (*Spartina alterniflora*) and native (*S. foliosa*) cordgrass (Poaceae) in California, USA determined by random amplified polymorphic DNA (RAPDs). *Molecular Ecology*, 8: 1179-1186.

Ayres, D. R., D. L. Smith, K. Zaremba, S. Klohr & D. R. Strong (2004). Spread of exotic cordgrasses × hybrids (*Spartina sp.*) in the tidal marshes of San Francisco Bay, California, USA. *Biological Invasions*, 6: 221-231.

Ayres D. R., Zaremba K., Sloop C. M., et al. (2008). Sexual reproduction of cordgrass hybrids (*Spartina foliosa* x *alterniflora*) invading tidal marshes in San Francisco Bay. *Diversity and Distributions*, 14: 187-195.

Baumel A., Ainouche M. L., Levasseur J. E. (2001). Molecular investigations in populations of *Spartina anglica* C.E. Hubbard (Poaceae) invading coastal Brittany (France). *Molecular Ecology*, 10: 1689–1701.

Baumel A., Ainouche M. L., Bayer R. J., Ainouche A. K., Misset M. T. (2002). Molecular phylogeny of hybridizing species from the genus *Spartina* Schreb. (Poaceae). Mol. *Phylogent. Evol.* 22: 303–314.

Bertness M. D. (1991). Zonation of *Spartina patens* and *Spartina alterniflora* in a New England salt marsh. *Ecology* 72:138-148.

Bertness M. D. & Callaway R. M. (1994). Positive interactions in communities. Trends in Ecology and Evolution, 9: 191–193.

Blum M. J., Christina M. S., Debra R. A. & Donald R. S. (2004). Characterization of microsatellite loci in *Spartina* species (Poaceae). *Molecular Ecology Notes*, 4: 39-42.

Bonilla-Warford C. M & Zedler J. B. (2002). Potential for using native plant species in stormwater wetlands. *Environmental Management*, 29(3): 385-394.

Boyer K.E. & Zedler J. B. (1998). Effect of nitrogen additions in the vertical structure of a constructed cordgrass marsh. *Ecological Applications*, 8: 692-705.

Bradley P. M. & Morris J. T. (1991). Relative importance of ion exclusion, secretion, and accumulation in *Spartina alterniflora* Loisel. *Journal of Experimental Botany*. 42(245):

1525-1532.

Brooker R. W. & Callaghan T. V. (1998). The balance between positive and negative plant interactions and its relationship to environmental gradients: a model. Oikos, 81: 196–207.

Broome S. W., W. W. Woodhouse & E. D. Seneca (1974). Propagation of smooth cordgrass, *Spartina alterniflora*, from seed in North Carolina. *Chesapeake Sci.* 15: 214-221.

Bruno J. F. (2000). Facilitation of cobble beach plant communities through habitat modification by *Spartina alterniflora*. *Ecology*, 81: 1179–1192.

Callaway J. C. & M. N. Josselyn (1992). The introduction and spread of smooth cordgrass (*Spartina alterniflora*) in south San Francisco Bay. *Estuaries*, 15: 218-226.

Callaway R. M. & Aschehoug E. T. (2000). Invasive plants versus their new and old neighbors: a mechanism for exotic invasion. *Science*, 290: 521-523.

Callaway R. M.& Pennings S. C. (2000). Facilitation may buffer competitive effects: indirect and diffuse interactions among salt marsh plants. *American Naturalist*, 156: 416-424.

Cavalieri, A. J. (1983). Proline and glycinebetaine accumulation by *Spartina alterniflora* Loisel. Response to NaCl and Nitrogen in a controlled environment.*Oecologia*, 57: 20-24.

Chambers R.M & Mozdzer T.J. (1998). Effect of salinity and sulfide of *Phragmites australis* and *Spartina alterniflora* in a tidal saltmarsh. *Aquatic Botany*, 62(3):161–189.

Chen J. Q. & C. H. Chung (1992). Isozyme studies on the ecotype differentiation of *S. alterniflora*. In Qin P. & Chung C. H. (eds), *Applied Studies on Spartina*. Ocean Science Press, Beijing, 110-116.

Chen Z Y, Li B, Zhong Y et al. (2004). Local competitive effects of introduced *Spartina alterniflora* on *Scirpus mariqueter* at Dongtan of Chongming Island, the Yangtze River estuary. *Hydrobiologia*, 528: 9~106.

Chung C. H., R. Z. Zhuo & G. W. Xu (2004). Creation of *Spartina* plantations for reclaiming Dongtai, China, tidal flats and offshore sands. *Ecological Engineering* 23: 135-150.

Chung H. C. (2006). Forty years of ecological engineering with *Spartina* plantations in China. *Ecological Engineering*, 27: 49-57.

Church G. L. (1940). Cytotaxonomic studies in the gramineae *Spartina, Andropogon* and *Panicum*. *American Journal of Botany*, 27: 263-271.

Crockett R. P. (1997). A historical perspective of glyphosate use to control *Spartina alterniflora* in Willapa Bay, p. 83-84 In: K. Patten (ed.). Proc. 2nd Int. *Spartina* Conf., Olympia, Washington April, 1997. Wash. State Univ., Olympia, Washington State, USA.

Daehler C.C & Strong D. R. (1994). Variable reproductive output among clones of *Spartina alterniflora* invading San Francisco Bay, Califonia: the influence of herbivory, pollination, and establishment site. *American Journal of Botany*, 81(3):307-313.

Daehler C.C. & Strong D.R. (1995). Impact of high herbivore densities on introduced smooth cordgrass, *Spartina alterniflora*, invading San Francisco Bay, California. *Estuaries*, 18, 409–417.

Daehler C. C. & Strong D. R. (1996). Status, prediction and prevention of introduced cordgrass *Spartina spp.* Invasions in Pacific estuaries, USA. *Biological Conservation*, 78: 51-58.

Daehler C. C. & Strong D. R. (1997). Hybridization between introduced smooth cordgrass (*Spartina alterniflora*; Poaceae) and native California cordgrass (*S. foliosa)* in San Francisco Bay, California, USA. *American Journal of Botany*, 84(5): 607-611.

Daehler C. C. (1998). Variation in self-fertility and the reproductive advantage of self-fertility for an invading plant (*Spartina alterniflora*). *Evolutionary Ecology* 12: 553-568.

Daehler C. C., C. K. Anttila, D. R. Ayres, D. R. Strong & J. P. Bailey (1999). Evolution of a new ecotype of *Spartina alterniflora* (Poaceae) in San Francisco Bay, California, USA.

American Journal of Botany, 86: 543-546.

Daehler C. C. (2003). Performance comparisons of co-occurring native and alien invasive plants: Implications for conservation and restoration. *Annual Review of Ecology Evolution and Systematics*, 34:183-211

Dai T. & Wiegert R. G. (1997). Ramet population dynamics and net aerial primary productivity of *Spartina alterniflora*. *Ecology*, 77(1): 276-288.

Davis H. G., Taylor C. M., Civille J. C., Strong D. R. (2004a). An Allee effect at the front of a plant invasion: *Spartina alterniflora* in a Pacific estuary. *Journal of Ecology*, 92, 321-327.

Davis H. G., Taylor C. M., Lambrinos J. G., et al. (2004b). Pollen limitation causes on Allee effect in a wind-pollinated invasive grass (*Spartina alterniflora*). *Proc. Natl. Acad. Sci. USA*, 101(38): 13804~13807.

Davis H. G. (2005). r-Selected traits in an invasive population. *Evolutionary Ecology*, 19: 255–274.

Delaune, R.D., Patrick Jr.,W.H., Buresen, R.J. (1978). Sedimentation determined by Cs137 dating in a rapidly accreting saltmarsh. *Nature*, 275, 532–533.

Deng Z. F., An S. Q., Zhi Y. B. & C. F. Zhou (2006). Preliminary studies on invasive model and outbreak mechanism of exotic species, *Spartina alterniflora* Loisel. *Acta Ecologica Sinica*, 26: 2678-2686.

Deng Z. F., An S. Q., Zhou C. F., Wang Z. S., Zhi Y. B., Wang Y. J., Shi S. H., Chen L. and Zhao C. J. (2007). Genetic structure and habitat selection of the tall form *Spartina alterniflora* Loisel. in China. *Hydrobiologia*, 583 (1): 195-204.

Deng Z. F., An S. Q., Zhao C. J., et al. (2008). Sediment burial stimulates the growth and invasion of *Spartina alterniflora* Loisel. *Estuarine, Coastal and Shelf Science*. 76: 818-826.

Donnelly J. P. & Bertness M. D. (2001). Rapid shoreward encroachment of salt marsh cordgrass in response to accelerated sea-level rise. *Proc. Natl. Acad. Sci*. USA., 98: 14218~14223.

Ebasco Environmental. (1993). Noxious Emergent Plant Environmental Impact Statement. Element C - No Action: Efficacy and Impacts. Final Report, submitted to Washington State Department of Ecology, Olympia.

Edwards K. R. & Mills K. P. (2005). Aboveground and belowground productivity of *Spartina alterniflora* (smooth cordgrass) in natural and created Louisiana satl marshes. *Estuaries*, 28(2): 252-265.

Ellstrand N.C. & Schierenbeck K.A. (2000). Hybridization as a stimulus for the evolution of invasiveness in plants? Proceeding of the National Academy of Science, USA 97, 7043–7050.

Fang X. B., Subudhi P. K., Venuto B. C. and Harrison S. A. (2004). Mode of pollination, pollen germination, and seed set in smooth cordgrass (*Spartina alterniflora*, Poaceae). *International Journal of Plant Sciences*, 165 (3): 395-401.

Feist B. E. & C. A. Simenstada (2000). Expansion rates and recruitment frequency of exotic smooth cordgrass, *Spartina alterniflora* (Loisel) colonizing unvegetated littoral flats in Willapa Bay, Washington. *Estuaries*, 23(2):267-274.

Ferrenberg S. M. & R. Denno (2003). Competition as a factor underlying the abundance of an uncommon phytophagous insect, the salt marsh planthopper *Delphacodes penedetecta*. *Ecological Entomology*, 28:58-66.

Foss S. (1992). *Spartina*: Threat to Washington's saltwater habitat. Washington State Department of Agriculture bulletin, Olympia.

Gallagher J.L., G.F. Somers, D.M. Grant, and D.M. Seliskar. (1988). Persistent differences in two forms of *Spartina alterniflora*: A common garden experiment. *Ecology*, 69: 1005-1008.

Gerard V. A. (1999).Positive interactions between cordgrass, *Spartina alterniflora*, and the brown alga, Ascophyllum nodosum ecad scorpioides, in a mid-Atlantic coast salt marsh. *J. Exp. Mar. Biol. Ecol.*, 239: 157-164.

Gray A.J., Marshal D.F. & Raybould A.F. (1991). A century of evolution in *Spartina anglica*. Advan. Ecol. Res., 21:1-62.

Gray A.J. (1992). Saltmarsh plant ecology: zonation and succession revisited. In: Allen, J.R.L. & Pyke, K. (eds.) *Saltmarshes: morphodynamics, conservation and engineering significance.* Cambridge University Press, Cambridge, pp. 63-79.

Grevstad F. S., Strong D R, Garcia-Rossi D, Switzer R W, Wecker M S. (2003). Biological control of *Spartina alterniflora* in Willapa Bay, Washington using the planthopper Prokelisia marginata: agent specificity and early results. *Biological Control*, 27:32-42.

Grevstad F. S. (2005). Simulating control strategies for a spatially structured and weed invasion: *Spartina alterniflora* (Loisel) in Pacific Coast estuaries. *Biological Invasions*, 7: 665-677.

Guan Y. J. & An S. Q. (2003). Ecological benefits and ecological consequences of exotic *Spartina* in China. *Journal of Nanjing Forestry University*, 95–101.

Hastings A., Cuddington K., Davies K. F., et al. (2005). The spatial spread of invasions: new developments in theory and evidence. *Ecology Letters*, 8: 91-101.

He W. S., Feagin R., Lu J. J., et al., (2007). Impacts of introduced *Spartina alterniflora* along an elevation gradient at the Jiuduansha Shoals in the Yangtze Estuary, suburban Shanghai, China. *Ecol. Engin.*, 29: 245-248.

Hedge P., Kriwoken L. K. & Patten K., (2003). A review of *Spartina* management in Washington State, US. *Journal of Aquatic Plant Management*, 41: 82-90.

Hitchcock, C.L., A.Cronquist, M. Ownbey, and J.W. Thompson (1969). Vascular plants of the Pacific Northwest. Part 1: Vascular cryptogams, gymnosperms, and monocotyledons. University of Washington Press, Seattle.

Huberty A. F. & Denno R. (2001). Elemental composition (N and P) of host plants differentially affects the performance of two herbivorous insects. An Entomological Odyssey of ESA, the ESA 2001 Annual Meeting, San Diego.

Huckle J. M., Marrs R. B. & Potter J. (2002). Interspecific and intraspecific interactions between salt marsh plants: integrating the effects of environmental factors and density on plant performance. *Oikos*, 96:307-319.

Levine J M, Brewer J S, Bertness M D. (1998). Nutrients, competition and plant zonation in a New England salt marsh. *Journal of Ecology*, 86: 285–292.

Liao C. Z., Luo Y. Q., Jiang L. F., et al. (2007). Invasion of *Spartina alterniflora* enhanced ecosystem carbon and nitrogen stocks in the Yangtze estuary,China. *Ecosystem*, 10: 1351-1361.

Liao C. Z., Luo Y. Q., Fang C. M., et al. (2008). Litter pool sizes, decomposition, and nitrogen dynamics in *Spartina alterniflora* -invaded and native coastal marshlands of the Yangtze Estuary. *Oecologia*, 10.1007/s00442-008-1007-0.

Luiting V. T., J. R. Cordell & A. M. Olson (1997). Does exotic *Spartina alterniflora* change benthic invertebrate assemblages? Proceedings of the Second International *Spartina* Conference. Washington State University, Olympia, 48-50.

Major W. W., Grue C. E., Grassley J. M., Conquest L. L. (2003). Mechanical and chemical control of smooth cordgrass in Willapa Bay, Washington. *Journal of Aquatic Plant Management,* 41:6-12.

Marchant C.J. (1970). Evolution in *Spartina* (Gramineae) IV. The cytology of *S. alterniflora*

Loisel. in North America. *Bot. J. Linn. Soc.*, 63: 321-326.

Maricle B.R. & Lee R.W. (2002). Aerenchyma development and oxygen transport in the estuarine cordgrasses *Spartina alterniflora* and *S. anglica*. *Aquatic Botany,* 74: 109-120.

Mendelssohn I. A. & N. L. Kuhn (2004). Sediment subsidy: effects on soil–plant responses in a rapidly submerging coastal salt marsh. *Ecol. Engin.*, 21:115-128.

Metcalfe, W.S., A.M. Ellison & M.D. Bertness (1986). Survivorship and spatial development of *Spartina alterniflora* Loisel. (Gramineae) seedlings in a New England salt marsh. *Annals of Botany* 58:249-258.

McMillan R. O., Armstrong D. A. & P. A. Dinnel (1995). Comparison of intertidal habitat use and growth rates of two northern Puget Sound cohorts of Dungeness crab, Cancer magister. *Estuaries*, 18(2): 390-398.

Mobberley, D.G. (1956). Taxonomy and distribution of the genus *Spartina*. Iowa State College. Journal of Science, 30: 471-574.

Mooring M. T., A.W. Cooper & E.D. Seneca (1971). Seed germination response and evidence for height ecophenes in *Spartina alterniflora* from North Carolina. *Amer. J. Bot.* 58: 48-55.

Morris J. T., Sundareshwar P. V., Nietch C. T., et al. (2002). Responses of coastal wetlands to rising sea level. *Ecology*, 83(10): 2869~2875.

Nielsen L.B., Finster K., Welsh D. T., Donelly A., Herbert R. A., Wit R. & Lomstein B. A. (2001). Sulphate reduction and nitrogen fixation rates associated with rppts, rhizomes and sediments from *Zostera noltii* and *Spartina maritima* meadows. *Environmental microbiology*, 3(1): 63-71.

O'Brien D. L. & D. W. Freshwater (1999). Genetic diversity within tall form *S. alterniflora* Loisel. along the Atlantic and Gulf Coasts of the United States. *Wetlands* 19: 352-358.

Padgett David E. & J. L. Brown (1999). Effects of drainage and soil organic contents on growth of *Spartina alterniflora* (Poaceae) in an artificial salt marsh mesocosm. *American Journa of Botany*, 86(5):697-702.

Partridge T.R. (1987). *Spartina* in New Zealand. *New Zealand Journal of Botany*. 25:567-576.

Patten K. (2002). Smooth cordgrass (*Spartina alterniflora*) control with Imazapyr. *Weed Technology*, 16: 826–832

Patten K. (2003). Persistence and non-target impact of Imazapyr associated with smooth cordgrass control in an estuary. *Journal Aquatic Plant Management*, 41:1-6.

Pennings S. C., Grant M. B. & Bertness M. D. (2005). Plant zonation in low-latitude salt marshes: disentangling the roles of flooding, salinity and competition. *Journal of Ecology*, 93: 159-167.

Pennings S. C.& Silliman B. R. (2005). Linking biogeography and community ecology: Latitudinal variation in plant-herbivore interaction strength. *Ecology*, 86: 2310-2319.

Pezeshki S. R. (1997). Photosynthesis and root growth in *Spartina alterniflora* in relation to root zone aeration. *Photosynthetica*, 34(1): 107-114.

Perkins E. J., W. J. Streever, E. Davis & H. L. Fredrickson (2002). Development of amplified fragment length polymorphism markers for *Spartina alterniflora*. *Aquatic Botany* 74: 85-95.

Piehler M. F., Currin C. A., Cassanova R. & Paerl H. W. (1998). Development and N2-fixing activity of the benthic microbial community in transplanted *Spartina alterniflora* marshes in North Carolina. *Restoration Ecology*, 6(3): 290-296.

Proffitt C. E., S. E. Travis & K. R. Edwards (2003). Genotype and elevation influence *Spartina alterniflora* colonization and growth in a created salt marsh. *Ecological Applications*, 13 (1): 180-192.

Qian Y. Q. & K. Ma (1995). Bio-techniques and bio-safety. *Journal of Natural Resources* 4:

322-331.

Qin P., M. D. Jin, & M. Xie (1985). Community biomass among the three ecotypes of *S. alterniflora* in Luoyuan Bay, Fujian. *Journal of Nanjing University Special Issue*: 226-236.

Qin P. & Chung C. H. (1992). *Applied studies on Spartina*. Beijing: Oceanic Press. 61-73, 105-108.

Qin P., Xie M. & Jiang Y. S. (1997). Estimation of the ecological economics benefits of two *Spartina alterniflora* plantations in North Jiangsu,China. *Ecol. Engin.*, 8: 5-17.

Raybould A. F., Gray A. J., Lawrence M. J. et al. (1991). The evolution of *Spartina anglica* C.E. Hubbard (Gramineae): origin and genetic variability. *Biol. J. Linn. Soc.*, 43, pp. 111-126.

Raybould A. F., Gray A. J. & Hornby D.D. (2000). Evolution and current status of the salt marshes grass, *Spartina anglica*, in the Solent. In: Collins M, Ansell K, eds. *Solent science – a review*. Amsterdam: Elsevier Science, 299–302.

Richards C. L., J. L. Hamrick, L. A. Donovan & R. Mauricio (2004). Unexpectedly high clonal diversity of two salt marsh perennials across a severe environmental gradient. *Ecology Letters*, 7: 1155-1162.

Richards C. L., S. C. Pennings & L. A. Donovan (2005). Habitat range and phenotypic variation in salt marsh plants. *Plant Ecology*, 176: 263-273.

Sanchez J. M., SanLeon D. G. & Izco J. (2001). Primary colonization of mudflat estuaries by *Spartina maritima* (Curtis) Fernald in Northwest Spain: vegetation structure and sediment accretion. *Aquatic Botany*, 69: 15-25.

Sayce K. (1988). Introduced cordgrass, *Spartina alterniflora* (Loisel.) in salt marshes and tidelands of Willapa Bay, Washington. U.S. Fish and Wildlife Service, Willapa Bay National Wildlife Refuge, Ilwaco, Washington. 70 p. FWSI-87058(TS).

Seliskar D. M., Gallagher J. L., Burdick D. M. & Mutz L. A. (2002). The regulation of ecosystem functions by ecotypic variation in the dominant plant: a *Spartina alterniflora* salt-marsh case study. *Journal of Ecology*, 90: 1-11.

Seneca E. D. & S.W. Broome (1972). Seedling response to photoperiod and temperature by smooth cordgrass, *Spartina alterniflora*, from Oregon Inlet, North Carolina. *Chesapeake Sci.*, 13: 212-235.

Seneca E. D. (1974). Germination and seedling response of Atlantic and Gulf coasts populations of *Spartina alterniflora. Amer. J. Bot.*, 61: 947-956.

Silliman B R, Layman C A, Geyer K, Zieman J C. (2004). Predation by the black-clawed mud crab, *Panopeus herbstii*, in Mid-Atlantic salt marshes: Further evidence for top-down control of marsh grass production. *Estuaries*, 27:188-196.

Silliman B. R. & Zieman J. C. (2001). Top-down control of *Spartina alterniflora* production by periwinkle grazing in a Virginia salt marsh. *Ecology*, 82: 2830-2845.

Simenstad C. A. & R. M. Thom (1995). *Spartina alterniflora* (smooth cordgrass) as an invasive halophyte in Pacific Northwest estuaries. *Hortus Northwest* 6:9-12, 38-40.

Smart R. & Barko J.W. (1982). Nitrogen nutrition and Salinity tolerance of *Distichlis spicata* and *Spartina alterniflora*. *Ecology*, 61:630-638.

Spartina Task Force. (1994). *Spartina* Management Program: Integrated Weed Management for Private Lands in Willapa Bay, Pacific County, Washington. Unpublished report prepared for the Noxious Weed Board and County Commissioners, Pacific County, Washington.

Stribling J.M. (1997). The relative importance of sulfate availability in the growth of *Spartina alterniflora* and *S. cynosuroides*. *Aquatic Botany*, 56(2):131-143.

Sun S. C., Zhu X B and Lü C. Q. (2004). Function assessment and ecological controls on alien species cordgrass. *Chinese Journal of Ecology*, 23 (3): 93-98.

Taylor C. M., Davis H. G., Civille J. C., Grevstad F. S. &Hastings A. (2004). Consequences of an Allee effect in the invasion of a Pacific estuary by *Spartina alterniflora. Ecology*, 85(12): 3254-3266.

Taylor C. M. & Hastings A. (2004). Finding optimal control strategies for invasive species: a density-structured model for *Spartina alterniflora. Journal of Applied Ecology*, 41: 1049-1057.

Thom R. M. (1992). Accretion rates of low intertidal salt marshes in the Pacific Northwest. *Wetlands*, 12:147-156.

Thompson J. D. (1991). The biology of an invasive plant: what makes *Spartina anglica* so successful? *Bioscience*, 41(6):393~400.

Travis S. E., C. E. Proffitt, R. C. Lowenfeld & T. M. Mitchell (2002). A comparative assessment of genetic diversity among differently-aged populations of *Spartina alterniflora* on restored versus natural wetlands. *Restoration Ecology*, 10 (1): 37-42.

Travis S. E., C. E. Proffitt & K. Ritland (2004). Population structure and inbreeding vary with successional stage in created Spartina alterniflora marshes. *Ecological Applications*, 14: 1189-1202.

Travis S. E. & M. W. Hester (2005). A space-for time substitution reveals the long-term decline in genotypic diversity of a widespread salt marsh plant, *Spartina alterniflora*, over a span of 1500 years. *Journal of Ecology*, 93: 417-430.

Tyler A. C., Mastronicola T. A. & McGlathery K. J. (2003). Nitrogen fixation and nitrogen limitation of primary production along a natural marsh chronosequence. *Oecologia*, 136, 431-438.

Tyler A. C., Lambrinos J. G.. & Groshpolz E. D. (2007). Nitrogen inputs promote the spread of an invasive marsh grass. *Ecological Applications*, 17(7): 1886-1898.

Valiela, I., J.M. Teal, and W.G. Deuser. (1978). The nature of growth forms in the salt marsh grass *Spartina alterniflora. Amer. Natur.*, 112:461-470.

Wang W., Zhang K. & Ru S. G. (2003). The status of *Spartina spp.* invasion in the world and the advances controlling over them. *Marine Sciences*, 27: 38-42.

Wang Q. & An S. Q. (2006). Invasive *Spartina alterniflora*: Biology, Ecology and Management. *Acta Phytotaxonomica Sinica,* 44 (5): 559–588.

Wang A. J., Gao S., Jia J. J. (2006). Impact of *Spartina alterniflora* on sedimentary and morphological evolution of tidal salt marshes of Jiangsu, China. *Acta Oceanologica Sinica,* 28: 93-100.

Wijte A. H. B. M.& Gallagher J. L. (1996a). Effect of oxygen availability and salinity on early life history stages of salt marsh plants. I. Different germination strategies of *Spartina alterniflora* and *Phragmites australis* (Poaceae). *American Journal of Botany*, 83(10): 1337-1342.

Wijte A. H. B. M. & Gallagher J. L. (1996b). Effect of oxygen availability and salinity on early life history stages of salt marsh plants. II. Early seedling development advantage of *Spartina alterniflora* over *Phragmites australis* (Poaceae). *American Journal of Botany*, 83(10): 1343-1350.

Xu B. S. (1998). *Flora of Shanghai*. Shanghai: Science and Technology Literature Publishing House.

Xu, G.W. & Zhuo, R.Z. (1985). Preliminary studies of introduced *Spartina alterniflora* Loisel in China (I). *Journal of Nanjing University Research Advance of Spartina*, 212-225.

Zhang R. S., Y. M. Shen & L. Y. Lu (2004). Formation of *Spartina alterniflora* salt marshes on the coast of Jiangsu Province, China. *Ecological Engineering* 23: 95-105.

Zhao C. J., Deng Z. F., An S. Q., et al. (2008). Effects of nitrogen availibility and competition on leaf characteristics of *Spartina alterniflora* and *Phragmites australis*. *J. Plant Ecol.*, 32: 392-401.

Zhao G. Q., Zhang L. Q. & Liang X. (2005). A comparison of photosynthetic characteristics between aninvasive plant *Spartina alterniflora* and an indigenous plant *Phragmites australis*. *Acta Ecol. Sinica*, 25(7): 1604~1611.

Zhu H. G., Qin P. & H. Wang (2004). Functional group classification and target species selection for Yancheng Nature Reserve, China. *Biodiversity and Conservation*, 13: 1335-1353.

In: Invasive Species: Detection, Impact and Control
Editors: C.P. Wilcox and R.B. Turpin ISBN 978-1-60692-252-1

Chapter 2

CURRENT TRENDS IN INVASIVE ASCIDIAN RESEARCH

***Stephan G. Bullard*[1] *and Mary R. Carman*[2]**

[1]University of Hartford, Hillyer College, West Hartford, CT 06117, USA
[2]Woods Hole Oceanographic Institution, Woods Hole, MA 02543, USA

ABSTRACT

Ascidians are common members of benthic marine communities. Due to their strong competitive abilities and their simple trophic requirements ascidians are highly invasive. Because they only need a hard surface for attachment and abundant particulate food to flourish, ascidians are easily introduced to new locations and can readily persist once established. Invasive ascidians often have considerable impact on invaded habitats. Not only can they affect benthic communities, but they also cause major problems for humans by overgrowing aquaculture equipment and organisms and by heavily fouling ships and man-made structures. Because ascidians have traditionally been of little direct economic value, much less is known about their biology than for other marine taxa (e.g., crustaceans, bivalves and teleosts). However, this is starting to change. Due to the impact invasive ascidians have had throughout the world in the past 25-30 years, ascidians have become the focus of significant scientific attention. Over the last few years a great deal of work has been conducted to learn more about ascidian ecology, to assess the impact of invasive ascidians on invaded systems, to prevent the spread of ascidians and to control them once they have become established in new areas. This review synthesizes the latest research on invasive ascidians and highlights areas for further study.

INTRODUCTION

Ascidians, known by the common name sea squirts, are ubiquitous members of benthic marine communities. They are sessile as adults and grow attached to natural and man-made hard surfaces (e.g., rocks, bivalve shells, docks, boat hulls, etc). Aside from a few predatory deep sea species, ascidians are filter feeders that mainly feed on phytoplankton and bacteria.

Ascidians can grow in a variety of forms ranging from purely solitary individuals to true colonies (Ruppert & Barnes 1996). All types of ascidians may be invasive, but the characteristics of their invasions differ. Invasive solitary species grow in dense monospecific aggregations (e.g., Bourque et al. 2003, Blum et al. 2007, LeBlanc et al. 2007, Figures 1, 2 & 3). Because each individual holds a significant amount of water inside its body, stands of solitary individuals can possess tremendous mass. For example, fouling clusters of *Ciona intestinalis* can weigh ~2 kg m^{-1} of line on mussel cultures and can add ~20 kg of mass to a single buoy (Daigle & Herbinger 2008, MacDonald *personal communication*). In contrast, invasive colonial species form blanket-like mats (e.g., Bullard et al. 2007a, Figures 4, 5 & 6). These mats are relatively thin, but can form impenetrable sheets that prevent exchanges of water, food and larvae between the benthos and water column and can prevent predators from foraging in benthic habitats (e.g., Lengyel et al. 2008).

Figure 1. Dense aggregation of the solitary ascidian *Ciona intestinalis* covering a mussel aquaculture cage in Prince Edward Island, Canada. These types of cages are traditionally pulled from the water by hand, but when heavily fouled with ascidians, as in this photograph, a crane is required to lift them. Photograph by Garth Arsenault.

For a comprehensive review of ascidian ecology and natural history see Lambert (2005a). However, certain aspects of ascidian biology need to be addressed here as they play an important role in invasions success. These traits include ascidian reproductive capabilities, competitiveness and defense mechanisms.

Ascidians reproduce sexually and produce short-lived tadpole larvae (Svane & Young 1989). Reproduction differs depending on growth form. Solitary ascidians have external fertilization and their eggs develop into larvae in the plankton. Colonial ascidians have internal fertilization and release well developed brooded larvae. Once hatched or released, tadpole larvae swim for minutes to days (e.g., Olson and McPherson 1987, Stoner 1990, Bingham & Young 1991, Marshall & Keough 2003). Given this short larval period, ascidians possess limited dispersal capabilities and generally have very localized recruitment (Grosberg 1987, Davis & Butler 1989, Havenhand 1991). Colonial ascidian colonies increase in size through the asexual production of zooids and some colonial species can also disperse asexually through fragmentation (Stoner 1989, Lambert 2005b, Bullard et al. 2007b). Fragmentation may be a particularly important ecological characteristic for invasive species because disturbance events (e.g., storms, boats impacts on the benthos, dredging, etc) can create viable fragments that are transported by currents and can reattachment to the bottom.

Figure 2. Heavy fouling of *Styela clava* on a mooring line, Great Harbor, Woods Hole, Massachusetts. Other ascidians present include *Botrylloides violaceus*, *Didemnum vexillum* and *Ciona intestinalis*. Photograph by Anne Goodwin.

Figure 3. Close up of *Styela clava* fouling; this photograph is of the same mooring line as Figure 2.

Figure 4. Dense mats of the colonial ascidian *Didemnum vexillum* on the bottom of Long Island Sound, Connecticut at a depth of 113 feet. Photograph by Dave Cohen.

Figure 5. Colonies of *Didemnum vexillum* encrusting pilings and rocks in Narragansett Bay, Rhode Island. The visible colonies are growing into the intertidal zone and are exposed during low tide. Photograph by Chris Deacutis.

Ascidians are strong spatial competitors that are frequently at the top of competitive hierarchies (Grosberg 1981, Nandakumar et al. 1993, Rajbanshi & Pederson 2007). They aggressively compete for space in several ways. First, ascidian life history characteristics facilitate rapid reproduction that allows larvae to quickly colonize open surfaces (e.g., Bourgue et al. 2007, Arsenault & Davidson 2008). Second, ascidians grow quickly to monopolize space and overgrow sessile species (Grosberg 1981, Russ 1982, Bullard et al. 2007a). Finally, they frequently possess noxious chemicals that they use to attack other species and prevent the attachment of epibionts and spatial competitors (Jackson & Buss 1975, Davis 1991, Teo & Ryland 1995, Morris et al. 2008). These advantages give invasive ascidians a significant edge when they reach new habitats and allow them to invade even highly diverse communities.

Ascidians frequently possess potent anti-predator defenses. Chemical defenses are common in ascidians and include the production of secondary metabolites and inorganic acids and the concentration of heavy metals (Stoecker 1980, Lindquist et al. 1992, Pisut & Pawlik

2002, Tarjuelo et al. 2002, Jimenez et al. 2003). In addition, many species possess thick, protective tunics (Ruppert & Barnes 1996). Given these defenses, invasive ascidians are often relatively immune to predation.

Figure 6. *Botrylloides violaceus* colonies overgrowing rocks and the algae *Ascophyllum nodosum*, Sandwich, Massachusetts. Photograph by Dann Blackwood.

INVASION HISTORY

Human-mediated transport of invasive marine species, including ascidians, has been occurring for centuries (Carlton & Hodder 1995, Carlton 2003). The rate of non-native species introductions greatly increased during the 20th century with the dramatic increase in marine travel and the globalization of world economies (Cohen & Carlton 1998). Within the last thirty years, ascidians have invaded numerous habitats worldwide, often with severe consequences.

Every continent except Antarctica has been invaded by non-native ascidians (Lambert & Lambert 1998, Hewitt et al. 2004, Robinson et al. 2005, Rocha & Kremer 2005, Abdul & Sivakumar 2007, Gittenberger 2007). While no ascidians are known to have been introduced to Antarctica (but see Lewis et al. 2005, Lewis et al. 2006), some species have spread from the Antarctic biogeographical province (e.g., Lambert 2004). In recent years ascidian introductions have been occurring with increasing frequency. For example, by 1998 thirteen species of non-native ascidians were established along the California coast, most of which (61%) had arrived within the last 15 years (Lambert & Lambert 1998). Additional species have since invaded (Lambert & Lambert 2003, Bullard et al. 2007a). Similarly, at least five ascidian species have invaded New England within the last thirty years (Whitlatch & Osman 1999, Steneck & Carlton 2001, Bullard et al. 2007a) and four species have invaded Prince Edward Island, Canada within the last eleven years (Ramsay et al. 2007). Worldwide, most coastal areas have experienced similar levels of invasion. Some native species have also increased in abundance within their normal biogeographic boundaries (e.g., Bak et al. 1996).

Ascidians are highly invasive as a group, but some species pose a greater invasion risk than others. In describing the environmental tolerances of ascidians in Spain, Naranjo et al. (1996) divided common Spanish species into three groups: regressive, transgressive and tolerant. These groups roughly correspond to invasiveness. Regressive species (not likely invasive) are largely confined to natural surfaces and cannot tolerate stress, transgressive species (potentially invasive) are pioneering species that are rarely dominant, and tolerant species (likely invasive) thrive in many settings and are very tolerant of stress. Ten out of twenty-seven Spanish species (37% overall) were considered tolerant (Naranjo et al. 1996). Thus, an exceptionally high proportion of ascidians have the strong potential to become invasive.

Some invasive ascidians have been particularly successful and have developed cosmopolitan distributions. For example, *Didemnum vexillum* (previously *Didemnum* sp. A; Stefaniak et al. 2008) has undergone massive population explosions in North American, New Zealand, Europe and Japan (Coutts 2002a, Bullard et al. 2007a, Minchin & Sides 2006, Lambert & Stefaniak 2008, Stefaniak et al. 2008). Many other species including *Styela clava* (Berman et al. 1992, Coutts & Forrest 2005, Davis & Davis 2007), *Ciona intestinalis* (McDonald 2004, Nydam & Harrison 2007, LeGresley et al. 2008) and *Botrylloides violaceus* (Yamaguchi 1975, Lambert & Sanamyan 2001, Steneck & Carlton 2001, Gittenberger 2007) have similar global distributions.

Prince Edward Island, Canada (PEI) provides a striking model of ascidian invasions and their impacts. Non-native ascidians began invading PEI waters in 1997. Since then a total of four species have arrived, *Styela clava* (1997), *Ciona intestinalis* (2004), and *Botrylloides violaceus* and *Botryllus schlosseri* (both appearing in 2005) (Boothroyd et al. 2002, Ramsay et al. 2007). There is great concern that a fifth species, *Didemnum vexillum*, will invade in the near future. These ascidians have had tremendous negative effects on PEI's shellfish aquaculture industry. By overgrowing aquaculture equipment and organisms ascidians have made it difficult to process overgrown shellfish and have added significant physical weight to aquaculture gear (e.g., cages, lines, buoys, etc; Figure 1). Prior to the invasions mussel aquaculture contributed >$36 million to PEI's GDP; though the overall economic effects of the invasions have not been tallied, non-native ascidians have increased costs associated with aquaculture due to the additional maintenance, labor and processing required by fouled cultures (Department of Fisheries and Oceans Canada 2006). In 2003 it was estimated that the control of *Styela clava* cost $0.115 per pound of shellfish harvested (ACRDP 2003). Costs have increased as additional species have arrived. Paradoxically it was likely the initial success of the PEI aquaculture industry that first led to the establishment of the ascidians. The benthic environment of coastal PEI is mostly muddy and has relatively little natural hard substrate. The explosive growth of aquaculture during the 1990s (Department of Agriculture, Fisheries and Aquaculture, PEI 2003), led to a massive increase in the amount of substrate available for ascidian colonization (Locke et al. 2007). Once ascidians became established, transfers of infested aquaculture organisms aided their spread (Locke et al. 2007). Development of effective, economically viable control measures has been a challenge. Despite intense individual, scientific and governmental control efforts ascidian populations in PEI and Atlantic Canada remain in bloom conditions.

Ecological and Economic Impacts

Invasive ascidians have significant impacts on marine communities, aquaculture and human economies. Ecologically, they can alter biodiversity and species richness and affect the growth and survivorship of aquacultured organisms. Economic impacts of invasive ascidians can be severe and at times may threaten aquaculture industries. Despite these problems, invasive ascidians occasionally provide ecological or economic benefits.

Invasive ascidians greatly affect community ecology. Due to their strong competitive abilities, invasive ascidians frequently overgrow and outcompete resident species and reduce the space available for settlement of resident larvae (Osman & Whitlatch 1995, Osman & Whitlatch 1999, Osman & Whitlatch 2007, Rajbanshi & Pederson 2007). As a result, invasive ascidians can decrease species richness and change the biodiversity of invaded habitats (but see Whitlatch et al. 1995). For example, in San Francisco Bay, invertebrate species richness was negatively correlated with *Ciona intestinalis* abundance and many species were missing or rare from *C. intestinalis* dominated substrates (Blum et al. 2007). In New England, invasive ascidians affected the biodiversity of subtidal systems because fewer species settled on the surface of invasive colonial ascidians compared to the surfaces of other organisms (Dijkstra et al. 2007b, but see Bullard et al. 2004). Invasive ascidians may also affect fundamental ecological processes by reducing the strength of benthic-pelagic linkages. For example, dense mats of *Didemnum vexillum* in New England prevent pelagic predators from foraging effectively from the benthos (Lengyel et al. 2008, Mercer & Whitlatch 2008). Changing environmental conditions, such as increasing surface sea water temperatures associated with climate change, can facilitate the growth and spread of invasive ascidians and may exacerbate these problems (Stachowicz et al. 2002, Agius 2007, Djikstra et al. 2007a).

Heavy fouling by invasive ascidians can harm aquaculture organisms (mainly bivalves) by competing with them for food or physically attaching to their shells. Both ascidians and bivalves are highly efficient filter-feeders. While ascidians and bivalves share the same food source, efficiency of particle removal differs; bivalves are selective, whereas ascidians are incapable of particle selection (Petersen & Riisgård 1992, Riisgård & Larsen 2000 & 2001, Ward & Shumway 2004). In addition, the retention efficiency of particles < 4 μm is greatly reduced in bivalves, whereas it is almost 100% in ascidians due to the indiscriminate nature of the mucus net feeding structure (Riisgård & Larsen 2000 & 2001, Ward & Shumway 2004). Thus, the intensity of food competition between ascidians and cultured bivalves can vary depending on conditions (Lesser et al. 1992, LeBlanc et al. 2003, Daigle & Herbinger 2008). The physical presence of dense ascidian aggregations on aquaculture gear may harm cultured bivalves by reducing seawater exchange between cages and the environment and limiting food availability. As a result of several of these processes, cultured mussels in Atlantic Canada grew less and decreased in condition as ascidian density on gear increased (Daigle & Herbinger 2008). Very heavy ascidian fouling on mussel lines (2 kg ascidian m^{-1}) also resulted in ~50% mussel mortality (Daigle & Herbinger 2008, Ramsay et al. 2008).

Invasive ascidians can cause extensive economic damage. Many costs are associated with losses to aquaculture from reduced harvests and increased production expenditures. For example, in Nova Scotia *Ciona intestinalis* fouling causes a loss of $1.86 ($C) m^{-1} of mussel line due to decreased mussel growth and increased mortality (Daigle & Herbinger 2008). In an effort to reduce these losses, shellfish farmers typically remove ascidians from their cultures. However, cleaning fouling from aquaculture gear is labor intensive and costly and may contribute 30% (or more) to the operational expenses of shellfish farmers (e.g., Claereboudt et al. 1994). For example, in Saldanha Bay, South Africa, mussel farmers spend the equivalent of ~$13,000 ($US) per year to keep their mussel ropes ascidian-free (Robinson et al. 2005). In 2003 it was estimated that control of ascidians on PEI (at the time predominantly *Styela clava*) cost ~$0.115 ($C) per pound of shellfish (ACRDP 2003). Given these economic impacts, invasive ascidian infestations can threaten regional aquaculture industries.

In addition to losses to aquaculture, large scale ascidian control efforts can also incur significant costs. When *Didemnum vexillum* invaded Shakespeare Bay, New Zealand cost-benefit analyses were conducted to assess the costs of different management strategies (Sinner & Coutts 2003, Coutts & Sinner 2004). Strategies ranged from doing nothing to attempting a complete, immediate eradication with costs ranging from ~$175,000-810,000 ($NZD) (Sinner & Coutts 2003). In the end, the actual costs incurred by an attempted eradication campaign were ~$650,000 ($NZD), but the efforts failed to eliminate the species (Coutts & Forrest 2007). Costs of invasive ascidians can also be incurred from monitoring and research efforts. For example, in 2005 the Canadian Government and PEI Aquaculture Industry allocated $1,000,000 ($C) to research and monitoring of invasive ascidians in Atlantic Canada (Department of Fisheries & Oceans Canada 2005).

Despite their potential problems, invasive ascidians can have positive ecological and economic impacts (e.g., Rodriguez 2006). In the intertidal zone in Chile the invasive ascidian *Pyura praeputialis* acts as an ecosystem engineer and increases species richness by forming dense intertidal aggregations (Castilla et al. 2000, Castilla et al. 2002, Castilla et al. 2004). On Georges Bank, New England two species of polychaetes use the invasive ascidian *Didemnum vexillum* as a predator refuge and have increased population sizes in sediments under *D. vexillum* mats (Lengyel et al. 2008). Filtering by large populations of invasive ascidians reduce phytoplankton abundances and improve turbidity levels in shallow bays (Petersen & Riisgård 1992). Invasive ascidians also provide direct benefits to humans. A few species, including *Styela clava*, are consumed by humans and might be profitably harvested (Bullard et al. 2006, Karney & Rhee 2008), potentially beneficial natural products have been isolated from many species (Haefner 2003, Newman & Cragg 2004, Blunt et al. 2006), and invasive ascidians may serve as bio-indicators of pollution because ascidian diversity appears to correlate with pollution levels (Carman et al. 2007).

Vectors of Ascidian Introductions

Numerous vectors facilitate the introduction, movement and spread of invasive ascidians. Due to the short-lived nature of ascidian larvae, most introductions likely occur when viable juveniles, adults or colony fragments are transported to new habitats. The most commonly cited transport mechanisms include shipping vectors (i.e., hull fouling and ballast water) and the movement of ascidians with aquaculture organisms and equipment; a more recently identified transport vector is the movement of epibiotic ascidians attached to mobile benthic organism such as crustaceans. Additional factors (e.g., abundant uncolonized artificial surfaces) may assist with the establishment of ascidians once they have been introduced to a region.

Because ascidian larvae generally remain in the plankton for minutes to hours (Svane & Young 1989, Lambert 2005a), few ascidian introductions are caused by natural larval dispersal. Although short-lived larvae significantly hinder long distance movement of ascidians, they provide an effective mechanism for local dispersal. Once ascidians have been introduced to a new location, high levels of local larval recruitment can allow them to quickly colonize surrounding surfaces and rapidly spread from their point of origin. High larval densities and recruitment rates have been documented in areas experiencing ascidian blooms. For example, *Styela clava* larvae have been found at concentrations of 560 m^{-3} in PEI, Canada (Bourque et al. 2007) and *Ciona intestinalis* had weekly recruitment levels of ~165 per 100 cm^2 in Nova Scotia, Canada (Howes et al. 2007).

Shipping vectors, especially hull fouling, likely account for most ascidian introductions (Minchin et al. 2006, Locke et al. 2007, Dodgshun et al. 2007, Darbyson et al. 2008b). Boat hulls provide an ideal hard substrate for colonization by benthic organisms, including ascidians (Darbyson et al. 2008a; Figure 7). Once attached, ascidians move with their host ship and can spread from the hull to benthic surfaces or release larvae into the environment (Minchin & Gollasch 2003). Ascidian hull fouling is common. For example, Godwin (2003) found seven ascidian species (all non-native) attached to boat hulls during a survey of only eight ocean-going vessels in Hawaii. Both recreational and commercial vessels may contribute to the spread of ascidians, but slow moving, rarely cleaned vessels are of most concern. This is demonstrated by the fact that heavily fouled barges have been implicated in several ascidian introductions (Coutts 2002a, Locke et al. 2007, Coutts & Forrest 2007). Ascidians may also be transported to new areas when small boats are trailered between locations. For example, approximately 90% of one-year-old *Styela clava* attached to boat hulls survived out of the water for 48 h in daytime temperatures of ~30 °C (Darbyson et al. 2008a), suggesting that they could easily survive short term trailering.

Figure 7. Removal of heavy *Didemnum vexillum* fouling from the bottom of a boat in Ireland. Originally published in Minchin & Sides (2006), *Aquatic Invasions*. Photograph by Damien Offer.

Ship ballast water may play a role in ascidian introductions (Carlton & Gellar 1993). Viable ascidian larvae have been found in ballast water samples (Chu et al. 1997) and ballast water likely assists with short range ascidian transport (Lavoie et al. 1999). However, given the brief lifespan of ascidian larvae, it is unlikely that ballast water can cause long distance ascidian introductions unless reproducing adults are carried inside the ballast tanks. It could be possible for ballast water to transport viable colony fragments over long distances (e.g., Lambert 2005b, Bullard et al. 2007b).

Invasive ascidians can be transported with infested aquaculture organisms (e.g., Naylor et al. 2001, Rocha & Baptista 2008). For example, because ascidians readily grow on bivalve shells, unless fouling organisms are removed prior to shipment, attached ascidians will be transported along with fouled bivalves. Dijkstra et al. (2007a) speculated that *Botrylloides violaceus* and *Didemnum vexillum* may have been introduced to the Gulf of Maine when Pacific Oysters (*Crassostrea gigas*) were imported for aquaculture. On a more regional scale, intra-province transfers of mussels are thought to have helped establish populations of *Styela clava*, *Botrylloides violaceus* and *Botryllus schlosseri* throughout PEI, Canada (Locke et al. 2007).

Natural flora and fauna also serve as transport vectors for invasive ascidians. For example, ascidians are transported on detached eelgrass blades and may be rafted many kilometers in this way (Worcester 1994). Another recently identified ascidian transport mechanism is movement on crustacean carapaces. Bernier et al. (2008) discovered *Botrylloides violaceus* and *Molgula* sp. growing on several species of crustaceans including Atlantic rock crabs (*Cancer irroratus*), northern lady crabs (*Ovalipes ocellatus*) and American lobsters (*Homarus americanus*). *Didemnum vexillum* has also recently been found on lobster carapaces and hermit crab-carried gastropod shells on Georges Bank (Valentine *personal communication*). Attached ascidian hitchhikers may be deposited into new benthic habitats with molts and abandoned shells and larvae and gametes are likely released as the crustaceans move. Such movement might be considerable as American lobsters are capable of moving 100s km over several years (Cambell & Stasko 1986).

Additional factors may enhance the establishment and spread of invasive ascidians. Invasive ascidians are often more abundant on artificial than natural surfaces (Glasby et al. 2007) and man-made surfaces (docks, pilings, boat hulls, buoys, aquaculture equipment, etc) have become ubiquitous in marine habitats (Glasby & Connell 1999, Connell 2001, Minchin et al. 2006). Artificial structures may aid the establishment of invasive ascidians in several ways. First, artificial structures provide novel surfaces that may negate any "home court" competitive advantages possessed by native species (e.g., Tyrrell & Byers 2007). Second, man-made structures provide additional physical space that can be colonized by invasive ascidians (Locke et al. 2007). Third, if artificial structures are not present, ascidians may be prevented from invading native communities either because they are less likely to reach the area (fewer docks would support less boat traffic and thus less risk from hull fouling) or because limited free space is available for their establishment. For example, in Martha's Vineyard, Massachusetts native species (scallops, eelgrass and algae) adjacent to artificial surfaces with invasive ascidians were all fouled by ascidians, while scallops and plants near artificial surfaces without invasive ascidians were ascidian-free (Carman et al., 2008a).

Prevention, Removal Treatments and Mitigation Efforts

A variety of mitigation techniques are available to help control invasive ascidians. Preventative measures that keep surfaces free of ascidians (e.g., antifouling paints, air drying, mechanical scraping, etc) may be most useful because they limit the spread of ascidians. Once invasive ascidians become established in a new location they can be almost impossible to eradicate. Various physical and chemical treatments are used to remove ascidians from substrates, but most fouling control methods are inefficient, costly, time consuming or harmful to non-target species.

Keeping surfaces ascidian-free (especially boat hulls, docks, and aquaculture equipment) helps prevent ascidian invasions. One of the most commonly used preventive methods is the application of commercially available antifouling paints. Antifouling paints are very effective at stopping fouling and greatly reduce ascidian settlement (Bellas 2005, Bellas 2006, Darbyson et al. 2008a). A significant limitation of antifouling paints is that many are toxic or have negative environmental impacts (e.g., Claisse & Alzieu 1993, Fernandez & Pinherio 2007, Konstantinou 2006), thus they cannot be used in sensitive biological areas or for aquaculture. This may be changing, as a few promising antifouling paints have been developed that are non-toxic and can be applied directly onto living shellfish (De Nys et al. 2004).

Periodic exposure to air also keeps surfaces ascidian-free. Ascidians cannot tolerate desiccation stress (Valentine et al. 2007, Darbyson et al. 2008a), so air drying will kill them. As a result, floating docks and boats have been removed from the water to help control ascidian outbreaks (Coutts & Forrest 2007). Removing artificial substrates has the added benefit of reducing the amount of hard substrate available for ascidian growth (e.g., Glasby et al. 2007, Locke et al. 2007). Air drying can also eliminate ascidians from aquaculture gear and shellfish, but care must be taken to ensure that shellfish can withstand the desiccation stress. These stresses vary depending on air temperature, relative humidity and size of individual animals (Katayama and Ikeda 1987), so it is important to determine emersion tolerances of ascidians and bivalves prior to air drying (Valentine et al. 2007).

Site-specific mechanisms may be used to keep surfaces ascidian-free. For example, commercial shellfish growers have experimented with novel antifouling methods. First, they have temporarily placed aquaculture bags at the mouth of rivers in the hope that low salinity levels would reduce fouling (e.g., Thiyagarajan & Qian 2003, Bullard & Whitlatch 2008). Second, they have placed aquaculture bags in eelgrass (*Zostera marina*) beds to see if antifouling compounds produced by the plants (e.g., Zimmerman et al. 1995, Newby et al. 2006) would prevent fouling. Results of these efforts have been mixed, and more study is warranted to fully assess their effectiveness.

Public education provides an important key to prevention. Boaters, harbormasters, and private dock owners are often avid stewards of the marine environment and can help monitor and control invasive ascidian populations. Before the public can take action, they must first be made aware of the problems posed by invasive ascidians and taught to identify common invaders. To this end, numerous groups have prepared illustrated pamphlets that describe invasive ascidians (Biosecurity New Zealand, British Columbia Shellfish Growers Association, Environment Canada, Fisheries and Oceans Canada, NOAA Sea Grant, etc). Education efforts encourage interested users to combat invasive ascidians (by ensuring that their boat hulls are scraped regularly or by notifying governmental or scientific agencies when new invasive species are spotted) and increase public awareness of the problem. Increased awareness is an integral step to legislative support efforts. Current legislative regulations vary; for example, removal of invasive ascidians is voluntary for some user groups (recreational boat owners and dock owners), but compulsory for others (commercial shellfish growers).

There are many methods to remove ascidians from natural and artificial substrates. Physical removal treatments include mechanical scrapping, pressure washing, smothering, air drying (including UV light exposure; e.g., Bingham & Reyns 1999, Bingham & Reitzel 2000), etc. The most commonly used technique is the mechanical removal of fouling organisms by scraping or pressure washing (Coutts 2006, Minchin et al. 2006). Mechanical removal of hull fouling can prevent the spread of ascidians between areas (e.g., Minchin et al. 2006, Minchin & Sides 2006), though during heavy infestations the disposal of a large amounts waste ascidians can pose an environmental challenge (Locke *personal communication*; Figure 7). Ascidians can be physically removed from aquacultured bivalves to increase shellfish growth rates and to make the final shellfish product visually appealing. While manual brushing is highly effective, it is inefficient and costly because each bivalve must be individually handled. Pressure washing is an easier and quicker method for removing ascidians from shellfish (Clancey & Hinton 2003). For example, in PEI, Canada the tunic of solitary ascidians fouling mussel lines are ruptured with high pressure water jets that do not harm the bivalves. A significant concern associated with scraping and pressure washing is the production of viable ascidian fragments. Because fragments of some colonial ascidians can survive and reattach (Bullard et al. 2007b), cleaning activities should be conducted on land if possible and detached colonial ascidians should not be returned to the water.

Smothering also eliminates ascidians. Smothering is accomplished by covering infested surfaces with materials that prevent water and light exchange and result in anoxic conditions. Although highly effective, smothering is nonselective and non-target species are killed along with ascidians (Coutts 2006). Coutts (2006) details smothering techniques aimed at eliminating *Didemnum vexillum* from Shakespeare Bay, New Zealand (see also Coutts & Forrest 2007). In this situation, polyethylene plastic sheets were wrapped around wharf pilings and seafloor areas were covered with geotextile fabric and buried with dredge spoil. All of these smothering techniques proved effective, but had limitations. Most *D. vexillum* colonies were eliminated from the wharves, although some survived due to ruptures in the plastic wrap and the inability of the covering to form watertight seals over rough bottom features. Similarly, some colonies survived on the seafloor near the seams of the geotextile fabric. Dredge spoil (~10 cm thick) eliminated all *D. vexillum* from level seafloor areas (Coutts 2006, Coutts & Forrest 2007), but proved ineffective on sloped seabed areas (Coutts & Forrest 2007).

Ascidians can be removed using chemical sprays or immersions in chemical baths. Caustic chemicals (e.g., acetic acid, sodium hypochlorite etc.) and non-caustic chemicals that are harmful to marine organisms (e.g., freshwater, steam) have been used. Due to the potential risks to non-target species, especially in aquaculture settings, care is needed in selecting appropriate chemical treatments. Perhaps the most environmentally friendly "chemical" is freshwater. Unfortunately, some ascidians can tolerate freshwater unless they are immersed in it for relatively long periods (Katayama and Ikeda 1987, Denny 2008) and it often difficult to gain access to large quantities of freshwater in the field. Acetic acid (i.e., vinegar) effectively removes ascidians (Carver et al. 2003, Locke et al. 2008), but can harm bivalves and other organisms (Forrest et al. 2007, Locke et al. 2008) and may leave some ascidians alive after exposure (Denny 2008). Sodium hypochlorite (chlorine bleach) proved very effective at removing *Didemnum vexillum* from green mussel seed (Denny 2008); exposing *D. vexillum* to 0.5% chlorine bleach for two minutes killed 100% of ascidians but left seed mussels relatively unaffected. Other caustic chemicals have been used (formalin, detergents, calcium hydroxide, sodium hydroxide, etc.) with varying results.

A final method of ascidian control involves the biological control of ascidians by other organisms. Some predators feed on ascidians (Lambert 2005a) and may be useful in controlling ascidian populations (Osman & Whitlatch 1999, Osman & Whitlatch 2004). For example, sea urchins consumed solitary ascidians attached to scallop shells (Ross et al 2004). However, biological controls are highly species-specific (e.g., Simoncini & Miller 2007) and in some cases predators do not appear able to control ascidians (Carman et al. 2008b). Diseases may also be used to control invasive species (e.g., Mutze et al 2008) and could be effective at controlling ascidians (e.g., Moiseeva et al. 2004). To date, there has been no directed effort to culture any pathogen specifically for use in ascidian control.

The controls described above are primarily designed for use against small-scale infestations. They can be of great value to individual users who want to control ascidian growth in localized areas (e.g., individual shellfish growers or boat owners), but when used independently have limited impact on larger-scale ascidian outbreaks. Coastal management efforts involving large-scale coordinated approaches are needed to prevent the arrival of ascidians or to control them once they are established (e.g., Locke & Smith 2008).

One of the most vigorous and well documented attempts to control a newly established ascidian involve the efforts to eliminate *Didemnum vexillum* from Whangamata Harbor, Shakespeare Bay, New Zealand (Coutts 2002a, Coutts 2006, Coutts & Forrest 2007). The species was introduced to the area in December of 2001 when a heavily fouled barge was moored in Whangamata Harbor. The ascidian's place of origin is unclear because the barge had traveled extensively since its launch in Australia in 1969. *D. vexillum* had obviously arrived with the barge because the species had not been seen in the area before the barge's arrival and its distribution was initially confined to the barge's hull and the seafloor directly beneath it. Once introduced, *D. vexillum* began to proliferate rapidly. In August 2002, an initial containment effort using underwater vacuums removed ~80% of the ascidian biomass from areas immediately surrounding the barge, but by this time the species had already infested ~40% of the pilings at a nearby shipping wharf (Coutts 2002b, Coutts & Forrest 2007). Eleven months later, the species was firmly established in the bay and was located on pilings, boat hulls, moorings, and natural surfaces. It had also spread to an aquaculture facility 35 km away due to movement of an infested aquaculture pontoon. In September 2003 an intensive eradication campaign was launched in a bid to eliminate *D. vexillum* from Shakespeare Bay. Mitigation efforts included: 1) removing infested boats from the water to physically remove and desiccate the ascidian; 2) treating infested boat hulls with high doses of chlorine; 3) wrapping wharf pilings with polyethylene to smother the ascidian; 4) smothering infested seabed areas with dredge spoil; and 5) smothering infested seabed areas with small pore-size geotextile fabric (Coutts & Forrest 2007). Despite significantly reducing local *D. vexillum* population sizes, these herculean efforts failed to eliminate the species from the system. *D. vexillum* continues to persist and spread throughout the region (Coutts & Forrest 2007).

The results from Shakespeare Bay are significant because they demonstrate that vigorous control methods can temporarily reduce the biomass of introduced ascidians but may not be able to entirely remove them. In this case, initial control efforts began seven months post-invasion (Coutts & Forrest 2007). While it could be argued that this response time was too slow, it is probably well above average for government agencies lacking pre-existing, in place mitigation strategies (i.e., most governmental agencies worldwide). Since it was impossible to eradicate an ascidian invader under these favorable conditions, few ascidian invaders may be able to be contained after they become established. It is therefore essential that at-risk regions develop detection and control plans so that when faced with a new ascidian invasion they can mount rapid, coordinated responses supported by previously established funding sources (e.g., Locke & Smith 2008).

CONCLUSION

There is a great deal more to be learned about invasive ascidians. In terms of basic biology, we need to gain a better understanding of factors that lead to the spread of invasive ascidians and the ecological impacts they have on invaded communities. In particular, more research should be focused on invasive ascidian dispersal ranges and rates, especially in the case of dispersal of colony fragments. Basic questions that remain unanswered include: What is the rate of fragment production? How long can fragments survive under different environmental conditions (e.g., in ballast water, free-floating in the water column, in benthic habitats, etc)? What proportion of fragments successfully reattach? How do fragment dispersal rates and distances compare to larval dispersal (Worcester 1994)? Additionally, detailed information about the types of ecological conditions that favor the growth of invasive ascidians would facilitate the development of predictive models of ascidian spread. For example, are there correlations between successful ascidian invasions and levels of nutrient enrichment, pollution, shoreline development or coastal disturbance? Do invasive ascidians have adaptive abilities that native species lack? More work is also needed to assess the biological and ecological impacts that invasive ascidians have on marine communities and aquacultured organisms. How much of an impact do invasive ascidians have on biodiversity? Can invasive ascidians change large scale ecosystem functions? Are fouling-related decreases in shellfish growth rates different depending on the type of aquaculture gear used (e.g., cages versus lines)? What are the mortalities rates for shellfish during heavy ascidian outbreaks? What densities of ascidians become prohibitively detrimental to the aquaculture industry?

Additional management efforts and strategies are also needed. It would be of great value to develop comprehensive datasets of regional ascidian diversity and abundance. Baseline datasets allow researchers to quickly determine whether newly observed ascidians are recently arrived invasive species or blooming populations of cryptogenic native species. This question is not merely academic as in some cases government agencies handle the control native species differently than invasive species. Active management responses may be delayed until the "residency status" of a newly observed ascidian can be determined, but by the time control efforts begin the species may be firmly established. This is exactly what happened when *Didemnum vexillum* first appeared in Shakespeare Bay, New Zealand (Coutts & Forrest 2007).

Dedicated monitoring programs in at-risk areas would also be helpful and would facilitate the discovery of newly arrived invasive ascidians (e.g., Cohen et al. 2005). Currently most ascidian invasions are detected either when an expert fortuitously discovers an unfamiliar species or, more commonly, when the invasive population has gotten so large that its presence is noticed by the general public. For example, *Didemnum vexillum* was first observed in Long Island Sound, USA by researchers at Groton, CT in 2000 (Bullard et al. 2007a). By the time it was noticed, the species was already well established at numerous sites along the Connecticut coastline. The following year extraordinarily large populations (100s m in extent) were discovered in deep areas of Long Island Sound (Figure 4). This was the first time these sites had been surveyed for ascidians, so the species had likely been there for some time, possibly years. Had a dedicated coastal monitoring system been in place the species may have been detected much earlier.

Perhaps the most important area of work is developing and refining, ecologically friendly methods to control invasive ascidians. At present, most control efforts remain localized in scope (e.g., scraping ascidians off individual boat hulls, power washing infected mussel lines, etc), though larger comprehensive control efforts have been attempted (e.g. Coutts & Forrest 2007). The "holy grail" of invasive ascidian management would be the development an effective, inexpensive, easy to use, non-toxic control method that targets ascidians but does not harm other organisms.

References

Abdul, J. A. H., & Sivakumar, V. (2007). Occurrence and distribution of ascidians in Vizhinjam Bay (southwest coast of India). *Journal of Experimental Marine Biology and Ecology, 342,* 189-190.

ACRDP (2003). Atlantic Canadian tunicate workshop proceedings: final report. *Aquaculture Collaborative Research and Development Program, Report no. MG-02-11-003.*

Agius, B. P. (2007). Spatial and temporal effects of pre-seeding plates with invasive ascidians: growth, recruitment and community composition. *Journal of Experimental Marine Biology and Ecology, 342,* 30-39.

Arsenault, G., & Davidson, J. (2008). Temporal and spatial development of an infestation of *Styela clava* on mussel farms in Malpeque Bay, Prince Edward Island, Canada. *Aquatic Invasions, in press.*

Bak, R. P. M., Lambrechts, D. Y. M., Joenje, M., Nieuwland, G., & Van Veghel, M. L. J. (1996). Long-term changes on coral reefs in booming populations of a competitive colonial ascidian. *Marine Ecology Progress Series, 133,* 303-306.

Bellas, J. (2005). Toxicity assessment of the antifouling compound zinc pyrithione using early developmental stages of the ascidian *Ciona intestinalis*. *Biofouling, 21,* 289-296.

Bellas, J. (2006). Comparative toxicity of alternative antifouling biocides on embryos and larvae of marine invertebrates. *Science of the Total Environment, 367,* 573-585.

Bernier, R. Y., Locke, A., & Hanson, J. M. (2008). Lobsters and crabs as potential vectors for tunicate dispersal in the southern Gulf of St. Lawrence, Canada. *Aquatic Invasions, in press.*

Berman, J., Harris, L., Lambert, W., Buttrick, M., & Dufresne, M. (1992). Recent invasions of the Gulf of Maine: three contrasting ecological histories. *Conservation Biology, 6*, 435-441.

Bingham, B. L., & Reitzel, A. M. (2000). Solar damage to the solitary ascidian, *Corella inflata*. *Journal of the Marine Biological Association of the UK, 80*, 515-521.

Bingham, B. L., & Reyns, N. B. (1999). Ultraviolet radiation and distribution of the solitary ascidian *Corella inflata* (Huntsman). *Biological Bulletin, 196*, 94-104.

Bingham, B. L., & Young, C. M. (1991). Larval behavior of the ascidian *Ecteinascidia turbinata* Herdman: an in-situ experimental study of the effects of swimming and dispersal. *Journal of Experimental Marine Biology and Ecology, 145,* 189-204.

Blum, J. C., Chang, A. L., Liljesthröm, M., Schenk, M. E., Steinberg, M. K., & Ruiz, G. M. (2007). The non-native solitary ascidian *Ciona intestinalis* (L.) depresses species richness. *Journal of Experimental Marine Biology and Ecology, 342,* 5-14.

Blunt, J. W., Copp, B. R., Munro, M. H. G., Northcote, P. T., & Prinsep, M. R. (2006). Marine natural products. *Natural Products Reports, 23*, 26-78.

Boothroyd, F. A., MacNair, N. G., Landry, T., Locke, A., & Davidson, T. J. (2002). Dealing with an aquatic invader: the clubbed tunicate (*Styela clava*) in Prince Edward Island waters. *Bulletin of the Aquaculture Association of Canada, 102*, 98-99.

Bourque, D., Davidson, J., MacNair, N. G., Arsenault, G., LeBlanc, A. R., Landry, T., & Miron, G. (2007). Reproduction and early life history of an invasive ascidian *Styela clava* Herdman in Prince Edward Island, Canada. *Journal of Experimental Marine Biology and Ecology, 342,* 78-84.

Bourque, D., Landry, T, Davidson, J., & McNair, N. (2003). Impact of an invasive tunicate in Atlantic Canada: recruitment and competition. *Journal of Shellfish Research, 22*, 320.

Bullard, S. G., Lambert, G., Carman, M. R., Byrnes, J., Whitlatch, R. B., Ruiz, G., Miller, R. J., Harris, L., Valentine, P. C., Collie, J. S., Pederson, J., McNaught, D. C., Cohen, A. N., Asch, R. G., Dijkstra, J., & Heinonen, K. (2007a). The colonial ascidian *Didemnum* sp. A: current distribution, basic biology and potential threat to marine communities of the northeast and west coasts of North America. *Journal of Experimental Marine Biology and Ecology, 342,* 99-108.

Bullard, S. G., Sedlack, B., Reinhardt, J. F., Litty, C., Gareau, K., & Whitlatch, R. B. (2007b). Fragmentation of colonial ascidians: differences in reattachment capability among species. *Journal of Experimental Marine Biology and Ecology, 342,* 166-168.

Bullard, S. G., & Whitlatch, R. B. (2008) *In situ* growth of the colonial ascidian *Didemnum vexillum* under different environmental conditions. *Aquatic Invasions, in press*.

Bullard, S. G., Whitlatch, R. B., & Osman, R. W. (2004). Checking the landing zone: do invertebrate larvae avoid settling near superior spatial competitors? *Marine Ecology Progress Series, 280,* 239-247.

Bullard, S. G., Whitlatch, R. B., Shumway, S. E., & Osman, R. W. (2006). Scientists crying "foul": sea squirts are invading Long Island Sound! *Wrack Lines, 5*, 2-7

Cambell, A., & Stasko, A. B. (1986). Movements of lobsters (*Homarus americanus*) tagged in the Bay of Fundy, Canada. *Marine Biology, 92*, 393-404.

Carlton, J. T. (2003). Community assembly and historical biogeography in the North Atlantic Ocean: the potential role of human-mediated dispersal vectors. *Hydrobiologia, 503*, 1-8.

Carlton, J. T., & Gellar, J. B. (1993). Ecological roulette: the global transport of nonindigenous marine organisms. *Science, 261*, 78-82.

Carlton, J. T., & Hodder, J. (1995). Biogeography and dispersal of coastal marine organisms: experimental studies on a replica of a 16th-century sailing vessel. *Marine Biology, 121*, 721-730.

Carman, M. R., Bullard, S. G., & Donnelly, J. P. (2007) Water quality, nitrogen pollution, and ascidian diversity in coastal waters of southern Massachusetts, USA. *Journal of Experimental Marine Biology and Ecology, 342,* 175-178.

Carman, M. R., Hoagland, K. E., Green-Beach, E., & Grunden, D. W. (2008a). Tunicate faunas of two North Atlantic-New England islands: Martha's Vineyard, Massachusetts and Block Island, Rhode Island. *Aquatic Invasions, in press.*

Carman, M. R., Allen, H. M., Tyrrell, M. C. (2008b). Limited value of the common periwinkle snail *Littorina littorea* as a biological control for the invasive tunicate *Didemnum vexillum*. *Aquatic Invasions, in press.*

Carver, C. E., Chisholm, A., & Mallet, A. L. (2003). Strategies to mitigate the effect of *Ciona intestinalis* (L.) biofouling on shellfish production. *Journal of Shellfish Research, 22,* 621-631.

Castilla, J. C., Collins, A. G., Meyer, C. P., Guiñez, R., & Lindberg, D. R. (2002). Recent introduction of the dominant tunicate, *Pyura praeputialis* (Urochordata, Pyuridae) to Antofagasta, Chile. *Molecular Ecology, 11,* 1579-1584.

Castilla, J. C., Guiñez, R., Alvarado, J. L., Pacheco, C., & Varas, M. (2000). Distribution, population structure, population biomass and morphological characteristics of the tunicate *Pyura stolonifera* in the Bay of Antofagasta, Chile. *Marine Ecology, 21,* 161–174.

Castilla, J. C, Lagos, N. A., & Cerda, M. (2004). Marine ecosystem engineering by the alien ascidian *Pyura praeputialis* on a mid-intertidal rocky shore. *Marine Ecology Progress Series, 268,* 119-130.

Chu, K. H., Tam, P. F., Fung, C. H., & Chen, Q. C. (1997). A biological survey of ballast water in container ships entering Hong Kong. *Hydrobiologia, 352,* 201-206.

Claereboudt, M. R., Bureau, D., Cote, J., & Himmelman, J. H. (1994). Fouling development and its effects on the growth of juvenile scallops (*Placopecten magellanicus*) in suspended culture. *Aquaculture, 121,* 327-342.

Claisse, D., & Alzieu, C. (1993). Copper contamination as a result of antifouling paint regulations? *Marine Pollution Bulletin, 26,* 395-397.

Clancey, L., & Hinton, R. (2003). Distribution of the tunicate, *Ciona intestinalis*, in Nova Scotia. *Nova Scotia Department of Fisheries and Agriculture, Aquaculture and Fisheries Pamphlet.*

Cohen, A. N., & Carlton, J. T. (1998). Accelerating invasion rate in a highly invaded estuary. *Science, 279,* 555-558.

Cohen, A. N., Harris, L. H., Bingham, B. L., Carlton, J. T., Chapman, J. W., Lambert, C. C., Lambert, G., Ljubenkov, J. C., Murray, S. N., Rao, L. C., Reardon, K. & Schwindt, E. (2005). Rapid assessment survey for exotic organisms in Southern California bays and harbors, and abundance in port and non-port areas. *Biological Invasions, 7,* 995-1002.

Connell, S. D. (2001). Urban structures as marine habitats: an experimental comparison of the composition and abundance of subtidal epibiota among pilings, pontoons and rocky reefs. *Marine Environmental Research, 52,* 115-125.

Coutts, A. D. M. (2002a). A biosecurity investigation of a barge in the Marlborough Sounds. *Cawthorn Report, vol 744.* Nelson, New Zealand, Cawthorn Institute.

Coutts, A. D. M. (2002b). The development of incursion response tools – underwater vacuum and filtering system trials. *Cawthorn Report, vol 755.* Nelson, New Zealand, Cawthorn Institute.

Coutts, A. D. M. (2006). An evaluation of incursion response tools for invasive species: a case study of *Didemnum vexillum* in the Marlborough Sounds. *Cawthorn Report, vol 1093.* Nelson, New Zealand, Cawthorn Institute.

Coutts, A. D. M., & Forrest, B. M. (2005). Evaluation of eradication tools for the clubbed tunicate *Styela clava*. *Cawthorn Report, vol 1110*. Nelson, New Zealand, Cawthorn Institute.

Coutts, A. D. M., & Forrest, B. M. (2007). Development and application of tools for incursion response: lessons learned from the management of the fouling pest *Didemnum vexillum*. *Journal of Experimental Marine Biology and Ecology, 342,* 154-162.

Coutts, A. D. M., & Sinner, J. (2004). An updated benefit-cost analysis of management options for *Didemnum vexillum* in Queen Charlotte Sound. *Cawthorn Report, vol 925*. Nelson, New Zealand, Cawthorn Institute.

Daigle, R. M., & Herbinger, C. M. (2008). Ecological interactions between the vase tunicate (*Ciona intestinalis*) and the farmed blue mussel (*Mytilus edulis*) in Nova Scotia, Canada. *Aquatic Invasions, in press.*

Darbyson, E., Hanson, J. M., Locke, A., & Willison, J. H. M. (2008a). Settlement and potential for transport of clubbed tunicate on boat hulls. *Aquatic Invasions, in press.*

Darbyson, E., Locke, A., Hanson, J. M., & Willison, J. H. M. (2008b). Marine boating habits and the spread of invasive species in the Gulf of St. Lawrence. *Aquatic Invasions, in press.*

Davis, A. R. (1991). Alkaloids and ascidian chemical defense: evidence for the ecological role of natural products from *Eudistoma olivaceum*. *Marine Biology, 111*, 375-379.

Davis, A. R., & Butler, A. J. (1989). Direct observations of larval dispersal in the colonial ascidian *Podoclavella moluccensis* Sluiter: evidence for closed populations. *Journal of Experimental Marine Biology and Ecology, 127,* 189-203.

Davis, M. H., & Davis, M. E. (2007). The distribution of *Styela clava* (Tunicata, Ascidiacea) in European waters. *Journal of Experimental Marine Biology and Ecology, 342,* 182-184.

De Nys, P. C., Steinberg, P. D., Charlton, T. S., & Christov, V. (2004). Antifouling of shellfish and aquaculture apparatus. *US Patent, no. 6,692,557.*

Denny, C. M. (2008). Development of a method to reduce the spread of *Didemnum vexillum* with aquaculture transfers. *ICES Journal of Marine Science*, 65.

Department of Agriculture, Fisheries and Aquaculture, PEI (2003). Mussel culture in Prince Edward Island. *Aqua Info Aquaculture Notes, Department of Agriculture, Fisheries and Aquaculture Prince Edward Island Canada.*

Department of Fisheries and Oceans Canada (2005). Governments of Canada and PEI announce funding for invasive tunicates. *Fisheries and Oceans Canada, Media Room, News Release.*

Department of Fisheries and Oceans Canada (2006). An economic analysis of the mussel industry in Prince Edward Island – Gulf Region. *Moncton, New Brunswick, Policy and Economic Branch, Gulf Region.*

Dijkstra, J., Harris, L. G., & Westerman, E. (2007a). Distribution and long-term patterns of four invasive colonial ascidians in the Gulf of Maine. *Journal of Experimental Marine Biology and Ecology, 342,* 61-68.

Dijkstra, J., Sherman, H., & Harris, L. G. (2007b). The role of colonial ascidians in altering biodiversity in marine fouling communities. *Journal of Experimental Marine Biology and Ecology, 342,* 169-171.

Dodgshun, T. J., Taylor, M. D., & Forrest, B. M. (2007). Human-mediated pathways of spread for non-indigenous marine species in New Zealand. *DOC Research & Development Series, no. 266.* Wellington, New Zealand, Science and Technical Publishing.

Fernandez, M. A., & Pinheiro, F. M. (2007). New approaches for monitoring the marine environment: the case of antifouling paints. *International Journal of Environment and Health, 1*, 427-448.

Forrest, B. M., Hopkins, G. A., Dodgshun, T. J., & Gardner, J. P. A. (2007). Efficacy of acetic acid treatments in the management of marine biofouling. *Aquaculture, 262*, 319-332.

Glasby, T. M., & Connell, S. D. (1999). Urban structures as marine habitats. *Ambio, 28,* 595-599.

Glasby, T. M., Connell, S. D., Holloway, M. G., & Hewitt, C. L. (2007). Nonindigenous biota on artificial structures: could habitat creation facilitate biological invasions? *Marine Biology, 151*, 887-895.

Gittenberger, A. (2007). Recent population expansions of non-native ascidians in The Netherlands. *Journal of Experimental Marine Biology and Ecology, 342,* 122-126.

Godwin, L. S. (2003). Hull fouling of maritime vessels as a pathway for marine species invasions to the Hawaiian Islands. *Biofouling, 19, Supplement 1*, 123-131.

Grosberg, R. K. (1981). Competitive ability influences habitat choice in marine invertebrates. *Nature, 290,* 700-702.

Grosberg, R. K. (1987). Limited dispersal and proximity-dependent mating success in the colonial ascidian *Botryllus schlosseri. Evolution, 41*, 372-384.

Haefner, B. (2003). Drugs from the deep: marine natural products as drug candidates. *Drug Discovery Today, 8*, 536-544.

Havenhand, J. N. (1991). Fertilisation and the potential for dispersal of gametes and larvae in the solitary ascidian *Ascidia mentula* Mueller. *Ophelia, 33*, 1-15.

Hewitt, C. L., Campbell, M. L., Thresher, R. E., Martin, R. B., Boyd, S., Cohen, B. F., Currie, D. R., Gomon, M. F., Keough, M. J., Lewis, J. A., Lockett, M. M., Mays, N., McArthur, M. A., O'Hara, T. D., Poore, G. C. B., Ross, D. J., Storey, M. J., Watson, J. E., & Wilson, R. S. (2004). Introduced and cryptogenic species in Port Phillip Bay, Victoria, Australia. *Marine Biology, 144*, 183-202.

Howes, S., Herbinger, C. M., Darnell, P., & Vercaemer, B. (2007). Spatial and temporal patterns of recruitment of the tunicate *Ciona intestinalis* on a mussel farm in Nova Scotia, Canada. *Journal of Experimental Marine Biology and Ecology, 342,* 85-92.

Jackson, J. B. C., & Buss, L. (1975). Allelopathy and spatial competition among coral reef invertebrates. *Proceedings of the National Academy of Sciences of the United States of America, 72*, 5160-5163.

Jimenez, P. C., Fortier, S. C., Lotufo, T. M. C., Pessoa, C., Moraes, M. E. A, de Moraes, M. O., & Costa-Lotufo, L.V. (2003). Biological activity in extracts of ascidians (Tunicata: Ascidiacea) from the northeastern Brazilian coast. *Journal of Experimental Marine Biology and Ecology, 287,* 93-101.

Karney, R. C., & Rhee, W. Y. (2008). Market potential for *Styela clava*, a non-indigenous pest invading New England coastal waters. *Aquatic Invasions, in press.*

Katayama, K., & Ikeda, Z. (1987). Tolerance of freshwater, hot water and sun-drying by *Didemnum moseleyi*, fouling organisms attached to culture oyster. *Bulletin of the Fisheries Experiment Station, Okayama Prefecture 2*, 104-106.

Konstantinou, I. K. (2006). Antifouling Paint Biocides. Berlin, Springer.

Lambert, C. C., & Lambert, G. (1998). Nonindigenous ascidians in southern California harbors and marinas. *Marine Biology, 130*, 675-688.

Lambert, C. C., & Lambert, G. (2003). Persistence and differential distribution of nonindigenous ascidians in harbors of the southern California Bight. *Marine Ecology Progress Series, 259*, 145-161.

Lambert, G. (2004). The south temperate and Antarctic ascidian *Corella eumyota* reported in two harbours in north-western France. *Journal of the Marine Biological Association of the UK, 84,* 239-241.

Lambert, G. (2005a). Ecology and natural history of the protochordates. *Canadian Journal of Zoology, 83*, 34-50.

Lambert, G. (2005b). First North American record of the ascidian *Perophora japonica. Journal of the Marine Biological Association of the UK, 85*, 1011-1012.

Lambert, G., & Sanamyan, K. (2001). *Distaplia alaskensis* sp. nov. (Ascidiacea, Aplousobranchia) and other new ascidian records from south-central Alaska, with a redescription of *Ascidia columbiana* (Huntsman, 1912). *Canadian Journal of Zoology, 79*, 1766-1781.

Lambert, G., & Stefaniak, L. (2008). Adventures of a sea squirt sleuth: unraveling the identity of *Didemnum vexillum*, a global ascidian invader. *Aquatic Invasions, in press*.

Lavoie, D. M., Smith, L. D., & Ruiz, G. M. (1999). The potential for intracoastal transfer of non-indigenous species in the ballast water of ships. *Estuarine, Coastal and Shelf Science, 48*, 551-564.

LeBlanc, A. R., Landry, T., & Miron, G. (2003). Fouling organisms of the blue mussel *Mytilus edulis*: their effect on nutrient uptake and release. *Journal of Shellfish Research, 22*, 633-638.

LeBlanc, N., Davidson, J., Tremblay, R., McNiven, M., & Landry, T. (2007). The effect of anti-fouling treatments for the clubbed tunicate on the blue mussel, *Mytilus edulis. Aquaculture, 264,* 205-213.

LeGresley, M. M., Martin, J. L., McCurdy, P., Thorpe, B., & Chang, B. D. (2008). Non-indigenous tunicate species in the Bay of Fundy, eastern Canada. *ICES Journal of Marine Science*.

Lengyel, N. L., Collie, J. S., & Valentine, P. C. (2008). The invasive colonial ascidian *Didemnum vexillum* on Georges Bank: ecological effects and genetic identification. *Aquatic Invasions, in press.*

Lesser, M. P., Shumway, S. E., Cucci, T., & Smith, J. (1992). Impact of fouling organisms on mussel rope culture: interspecific competition for food among suspension-feeding invertebrates. *Journal of Experimental Marine Biology and Ecology, 165*, 91-102.

Lewis, P. N., Bergstrom, D. M., & Whinam, J. (2006). Barging in: a temperate marine community travels to the subantarctic. *Biological Invasions, 8*, 787-795.

Lewis, P. N., Riddle, M. J., & Smith, S. D. A. (2005). Assisted passage or passive drift: a comparison of alternative transport mechanisms for non-indigenous coastal species into the Southern Ocean. *Antarctic Science, 17*, 183-191.

Lindquist, N., Hay, M. E., & Fenical, W. (1992). Defense of ascidians and their conspicuous larvae: adult vs. larval chemical defense? *Ecological Monographs, 62*, 547-568.

Locke, A., & Smith A. H. (2008). Rapid response to invasive tunicates in Atlantic Canada: a proposed framework and case studies from Prince Edward Island. *Aquatic Invasions, in press.*

Locke, A., Doe, K. G., Fairchild, W. L., Jackman, P. M., & Reese, E. J. (2008). Effects of tunicate management with acetic acid and calcium hydroxide on non-target marine organisms in Prince Edward Island, Canada. *Aquatic Invasions, in press.*

Locke, A., Hanson, J. M., Ellis, K. M., Thompson, J., & Rochette, R. (2007). Invasion of the southern Gulf of St. Lawrence by the clubbed tunicate (*Steyla clava* Herdman): potential mechanisms for invasions of Prince Edward Island estuaries. *Journal of Experimental Marine Biology and Ecology, 342*, 69-77.

Marshall, D. J., & Keough, M. J. (2003). Variation in the dispersal potential of non-feeding invertebrate larvae: the desperate larva hypothesis and larval size. *Marine Ecology Progress Series, 255*, 145-153.

McDonald, J. (2004). The invasive pest species *Ciona intestinalis* (Linnaeus, 1767) reported in a harbour in southern Western Australia. *Marine Pollution Bulletin, 49*, 868-870.

Mercer, J., & Whitlatch, R. B. (2008) Potential effects of the invasive colonial ascidian (*Didemnum vexillum*) on pebble-cobble bottom habitats in southern New England, USA. *Aquatic Invasions, in press*.

Minchin, D., & Gollasch, S. (2003). Fouling and ships' hulls: how changing circumstances and spawning events may result in the spread of exotic species. *Biofouling, 19, Supplement 1*, 111-122.

Minchin, D., & Sides, E. (2006). Appearance of a cryptogenic tunicate, a *Didemnum* sp. fouling marina pontoons and leisure craft in Ireland. *Aquatic Invasions, 1*, 143-147.

Minchin, D., Floerl, O., Savini, D., & Occhipinti-Ambrogi, A. (2006). Small craft and the spread of exotic species. In J. Davenport & J. L. Davenport (Eds.), *The Ecology of Transportation: Managing Mobility for the Environment* (pp. 99-118). The Netherlands, Springer.

Moiseeva, E., Rabinowitz, C., Yankelevich, I., & Rinkevich, B. (2004). 'Cup cell disease' in the colonial tunicate *Botryllus schlosseri*. *Diseases of Aquatic Organisms, 60*, 77-84.

Morris, J., Carman, M. R., Hoagland, K. E., Green-Beach, E., & Karney, R. (2008). Impacts of the invasive colonial tunicate *Didemnum vexillum* on the recruitment of the bay scallop (*Argopecten irradians irradians*) and implications for recruitment of the sea scallop (*Placopecten magellanicus*) on Georges Bank. *Aquatic Invasions, in press*.

Mutze, G., Bird, P., Cooke, B., & Henzell, R. (2008). Geographic and seasonal variation in the impact of rabbit haemorrhagic disease on European rabbits, *Oryctolagus cuniculus*, and rabbit damage in Australia. In P. C. Alves, N. Ferrand, & K. Hackländer (Eds.), *Lagomorph Biology Evolution, Ecology, and Conservation* (pp. 279-293). Berlin, Springer.

Nandakumar, K., Tanaka, M., & Kikuchi, T. (1993). Interspecific competition among fouling organisms in Tomioka Bay, Japan. *Marine Ecology Progress Series, 94,* 43-50.

Naranjo, S. A., Carballo, J. L., & García-Gómez, J. C. (1996). Effects of environmental stress on ascidian populations in Algeciras Bay (southern Spain). Possible marine bioindicators? *Marine Ecology Progress Series, 144*, 119-131.

Naylor, R. L., Williams, S. L., & Strong, D. R. (2001). Aquaculture - a gateway for exotic species. *Science, 294*, 1655-1656.

Newby, B. Z., Cutright, T., Barrios, C. A., & Xu. Q. (2006). Zosteric acid—An effective antifoulant for reducing freshwater bacterial attachment on coatings. *Journal of Coatings Technology and Research, 3*, 69-76.

Newman, D. J., & Cragg, G. M. (2004). Marine natural products and related compounds in clinical and advanced preclinical trials. *Journal of Natural Products, 67*, 1216-1238.

Nydam, M. L., & Harrison, R. G. (2007). Genealogical relationships within and among shallow-water *Ciona* species (Ascidiacea). *Marine Biology, 151*, 1839-1847.

Olson, R. R., & McPherson, R. (1987). Potential vs. realized dispersal: fish predation on larvae of the ascidian *Lissoclinum patella* (Gottschaldt). *Journal of Experimental Marine Biology and Ecology, 110,* 245-256.

Osman, R. W., & Whitlatch, R. B. (1995). Influence of resident adults on recruitment: a comparison to settlement. *Journal of Experimental Marine Biology and Ecology, 190,* 169-198.

Osman, R. W., & Whitlatch, R. B. (1999). Ecological interactions of invading ascidians within epifaunal communities of southern New England. In J. Pederson (Ed.), *Marine Bioinvasions, Proceedings of the Northeast Conference on Non-Indigenous Aquatic Nuisance Species: a Regional Conference* (pp. 164-174). Cambridge: Massachusetts Institute of Technology.

Osman, R. W., & Whitlatch, R. B. (2004). The control of the development of a marine benthic community by predation on recruits. *Journal of Experimental Marine Biology and Ecology, 311,* 117-145.

Osman, R. W., & Whitlatch, R. B. (2007). Variability in the ability of *Didemnum* sp. to invade established communities. *Journal of Experimental Marine Biology and Ecology, 342,* 40-53.

Petersen, J. K., & Riisgård, H. U. (1992). Filtration capacity of the ascidian *Ciona intestinalis* and its grazing impact in a shallow fjord. *Marine Ecology Progress Series, 88*, 9-17.

Pisut, D. P., & Pawlik, J. R. (2002). Anti-predatory chemical defenses of ascidians: secondary metabolites or inorganic acids? *Journal of Experimental Marine Biology and Ecology, 270,* 203-214.

Rajbanshi, R., & Pederson, J. (2007). Competition among invading ascidians and a native mussel. *Journal of Experimental Marine Biology and Ecology, 342,* 163-165.

Ramsay, A., Davidson, J., Landry, T., & Arsenault, G. (2007). Process of invasiveness among exotic tunicates in Prince Edward Island, Canada. *Biological Invasions*.

Ramsay, A., Davidson, J., Landry, T., & Stryhn, H. (2008). The effect of mussel seed density on tunicate settlement and growth of the cultured mussel, *Mytilus edulis*. *Aquaculture, 275*, 194-200.

Riisgård, H. U., & Larsen, P. S. (2000). Comparative ecophysiology of active zoobenthic filter feeding, essence of current knowledge. *Journal of Sea Research 44*:169–193.

Riisgård, H. U., & Larsen, P. S. (2001). Minireview: ciliary filter feeding and bio-fluid mechanics—present understanding and unsolved problems. *Limnology and Oceanography, 46*, 882-891.

Robinson, T. B., Griffiths, C. L., McQuaid, C. D., & Ruis, M. (2005). Marine alien species of South Africa – status and impacts. *African Journal of Marine Science, 27,* 297-306.

Rocha, R. M., & Baptista, M. S. (2008). Invasive tunicates in oyster cultivation: potential to colonize the natural substrate. *Aquatic Invasions, in press*.

Rocha, R. M., & Kremer, L. P. (2005). Introduced ascidians in Paranaguá Bay, Paraná, southern Brazil. *Revista Brasileira de Zoologia, 22*, 1170-1184.

Rodriguez, L. F. (2006). Can invasive species facilitate native species? Evidence of how, when, and why these impacts occur. *Biological Invasions, 8*, 927-939.

Ross, K. A., Thorpe, J. P., & Brand, A. R. (2004). Biological control of fouling in suspended scallop cultivation. *Aquaculture, 229*, 99-116.

Ruppert, E. E., & Barnes, R. D. (1996). *Invertebrate Zoology* (6th edition). Orlando, Saunders College Publishing.

Russ, G. R. (1982). Overgrowth in a marine epifaunal community: competitive hierarchies and competitive networks. *Oecologia, 53*, 12-19.

Simoncini, M., & Miller, R. J. (2007). Feeding preference of *Strongylocentrotus droebachiensis* (Echinoidea) for a dominant native ascidian, *Aplidium glabrum*, relative to the invasive ascidian *Botrylloides violaceus*. *Journal of Experimental Marine Biology and Ecology, 342,* 93-98.

Sinner, J., & Coutts, A. D. M. (2003). Benefit-cost analysis of management options for *Didemnum vexillum* in Shakespeare Bay. *Cawthorn Report, vol 924*. Nelson, New Zealand, Cawthorn Institute.

Stachowicz, J. J., Terwin, J. R., Whitlatch, R. B., & Osman, R. W. (2002). Linking climate change and biological invasions: ocean warming facilitates nonindigenous species invasions. *Proceedings of the National Academy of Sciences USA, 99*, 15497-15500.

Steneck, R. S., & Carlton, J. T. (2001). Human alterations of marine communities: students beware! In M. D. Bertness, S. D. Gaines, & M. E. Hay, (Eds.), *Marine Community Ecology* (pp 445-468). Sunderland, Sinauer Associates.

Stefaniak, L., Lambert, G., Gittenberger, A., Zhang, H., Lin, S., & Whitlatch, R. B. (2008). Genetic conspecificity of the worldwide populations of *Didemnum vexillum* Kott, 2002. *Aquatic Invasions, in press*.

Stoecker, D. (1980). Relationships between chemical defense and ecology of benthic ascidians. *Marine Ecology Progress Series, 3*, 257-265.

Stoner, D. S. (1989). Fragmentation: a mechanism for the stimulation of genet growth rates in an encrusting colonial ascidian. *Bulletin of Marine Science, 45*, 277-287.

Stoner, D. S. (1990). Recruitment of a tropical ascidian: relative importance of pre-settlement vs. post-settlement processes. *Ecology, 71*, 1682-1690.

Svane, I., & Young, C. M. (1989). The ecology and behaviour of ascidian larvae. *Oceanography and Marine Biology Annual Review, 27*, 45-90.

Tarjuelo, I., López-Legentil, S., Codina, M., & Turon, X. (2002). Defence mechanisms of adults and larvae of colonial ascidians: patterns of palatability and toxicity. *Marine Ecology Progress Series, 235*, 103-115.

Thiyagarajan, V., & Qian, P.-Y. (2003). Effect of temperature, salinity and delayed attachment on development of the solitary ascidian *Styela plicata* (Lesueur). *Journal of Experimental Marine Biology and Ecology, 290*, 133-146.

Teo, S. L. M., & Ryland, J. S. (1995). Potential antifouling mechanisms using toxic chemicals in some British ascidians. *Journal of Experimental Marine Biology and Ecology, 188*, 49-62.

Tyrrell, M. C., & Byers, J. E. (2007). Do artificial substrates favor nonindigenous fouling species over native species? *Journal of Experimental Marine Biology and Ecology, 342*, 54-60.

Valentine, P. C., Carman, M. R., Blackwood, D. S., & Heffon, E. J. (2007). Ecological observations on the colonial ascidian *Didemnum* sp. in a New England tide pool habitat. *Journal of Experimental Marine Biology and Ecology, 342*, 109-121.

Ward, J. E., & Shumway, S. E. (2004). Separating the grain from the chaff: particle selection in suspension- and deposit-feeding bivalves. *Journal of Experimental Marine Biology and Ecology, 300*, 83-130.

Whitlatch, R. B., & Osman, R. W. (1999). Geographical distributions and organism-habitat associations of shallow-water introduced marine fauna in New England. In J. Pederson (Ed.), *Marine Bioinvasions, Proceedings of the Northeast Conference on Non-Indigenous Aquatic Nuisance Species: a Regional Conference* (pp. 61-65). Cambridge: Massachusetts Institute of Technology.

Whitlatch, R. B., Osman, R. W., Frese, A., Malatesta, R., Mitchell, P., & Sedgwick, L. (1995). The ecology of two introduced marine ascidians and their effects on epifaunal organisms in Long Island Sound. In N. C. Balcolm (Ed.), *Proceedings of the Northeast Conference on Non-Indigenous Aquatic Nuisance Species* (pp. 29-48). Groton, Connecticut Sea Grant College Program.

Worcester, S. E. (1994). Adult rafting versus larval swimming: dispersal and recruitment of a botryllid ascidian on eelgrass. *Marine Biology, 121*, 309-317.

Yamaguchi, M. (1975). Growth and reproductive cycles of the marine fouling ascidians *Ciona intestinalis, Styela plicata, Botrylloides violaceus*, and *Leptoclinum mitsukurii* at Aburatsubo-Moroiso Inlet (central Japan). *Marine Biology, 29*, 253-259.

Zimmerman, R. C., Alberte, R. S., Todd, J. S., & Crews P. (1995). Phenolic acid sulfate esters for prevention of marine biofouling. *US Patent, no. 5,384,176*.

In: Invasive Species: Detection, Impact and Control
Editors: C.P. Wilcox and R.B. Turpin
ISBN 978-1-60692-252-1

Chapter 3

PREVENTION: A PROACTIVE APPROACH TO THE CONTROL OF INVASIVE PLANTS IN WILDLANDS

Kirk W. Davies[1,*] and Dustin D. Johnson[2]

[1]Rangeland Scientist, USDA – Agricultural Research Service, Eastern Oregon Agricultural Research Center, Burns, Oregon, USA

[2]Assistant Professor, Department of Range Ecology and Management, Harney County Extension Office, Oregon State University, Burns, Oregon, USA

ABSTRACT

Infestations of wildlands by invasive plants can reduce resource productivity, decrease biodiversity, displace native vegetation, and alter ecosystem processes and functions. The traditional reactive strategy of controlling established invasive plant infestations followed by restoration of the native plant community has proven to be largely ineffective at reducing the spread and negative impacts of invasive plants. This approach often fails in its attempt to restore native plant communities and is too costly to apply at the scale required to have substantial effects. While large amounts of resources are intensively spent on efforts to restore a few infested wildlands, invasive plants continue to spread via emerging populations and expanding established infestations. A proactive approach with the objective of preventing new infestations and limiting the expansion of existing infestations is a more effective and efficient strategy for managing invasive plants in wildlands because it precludes the need for restoration. However, relatively few resources are being directed towards preventing the spread of invasive species. Successful strategies to prevent infestations of invasive plants should focus on: 1) limiting the spatial dispersal of propagules (i.e., reducing propagule pressure), 2) maintaining or increasing the ability of wildland plant communities to resist invasion (i.e., biotic resistance), and 3) systematically searching for and eradicating new infestations. Propagule pressure and biotic resistance interact to determine wildland plant community invasibility. At low biotic resistance even a few propagules may result in successful invasion; however, as biotic resistance increases, greater propagule pressure is

* Correspondence: Kirk W. Davies, USDA-Agricultural Research Service, Eastern Oregon Agricultural Research Center, 67826-A Hwy 205, Burns, Oregon, 97720. Email: kirk.davies@oregonstate.edu Tel: (541) 573-4074

required for invasion. Thus, efforts aimed at decreasing invasive plant propagule pressure and increasing biotic resistance can greatly reduce new infestations. Systematically searching for and eradicating new infestations is also a critical element of a successful prevention strategy, because uncontrollable events may still lead to new infestations. Successful management of invasive plants in wildlands will require more efforts and resources directed at prevention. This task can be facilitated by more research developing and improving prevention strategies and demonstrating the effectiveness of proactive management.

INTRODUCTION

Invasive plants negatively impact wildlands by decreasing biodiversity, reducing production, displacing native species, degrading systems, and altering ecosystem functions and processes (DiTomaso 2000; Masters and Sheley 2001; Davies and Svejcar *IN PRESS*). Invasive species are the second leading threat to biodiversity after habitat destruction (Wittenberg and Cock 2001; Randall 1996; Pimm and Gilpin 1989). The negative impacts of invasive plants are a product of their ability to out-compete native vegetation for resources and/or alter conditions to facilitate their dominance to the detriment of native plant species. For example, invasive annual grasses, such as *Bromus tectorum* L. (cheatgrass) and *Taeniatherum caput-medusae* (L.) Nevski (medusahead), compete effectively with native vegetation for resources (Hironaka and Sindelar 1975; Mack 1981; Goebel et al. 1988; Young and Allen 1997; Young and Mangold 2008) and increase fire frequency to the detriment of native vegetation (Stewart and Hull 1949; Torell et al. 1961; Whisenant 1990; Young 1992) (Fig. 1). The negative economic impacts of invasive plants species were estimated to be $13 billion a year in the United States in 1994 (Westbrooks 1998). Though not solely focusing on invasive plant species, Pimentel et al. (2000) estimated that nonindigenous species caused environmental degradation and economical losses of about $137 billion in 2000. However, the rapid increase in the area infested with invasive plant species has also escalated the negative impacts of these invaders. Thus, the economic impacts of invasive plant species has probably increased substantially and multiplied every year as invasive species continue to spread and increase their dominance.

The traditional approach to managing invasive plant species on wildlands has been to control already established infestations followed by efforts to restore native vegetation. This reactive approach is expensive and often fails to remove the invasive plant species and/or restore displaced native plant species. Once invasive plant species have established rapidly spreading populations, eradication is rarely an option (Mack et al. 2000; Eiswerth and Johnson 2002). Eradication of invasive species can also produce negative secondary effects if the invaders have filled the functional roles of native species in the ecosystem (Zavaleta et al. 2001). Once an invasive species has established rapidly spreading populations, even the control of the invader is a significant sink of resources and time (Huenneke 1996). Control is often temporary and the invader often reestablishes dominance within a few growing seasons because restoration of native vegetation fails.

Restoration of native vegetation in wildlands infested with invasive plants is rarely successful and prohibitively expensive (Vitousek et al. 1997). Restoration is exceedingly difficult because invasive plants can leave a negative legacy of physical or chemical alterations to the site or a persistent invasive plant seed bank (D'Antonio and Meyerson

2002). These alterations may make it improbable that the native plant community can reclaim the site, even if the invasive species is eradicated (Cronk and Fuller 1995; D´Antonio and Meyerson 2002). Steep and rocky terrain common to wildlands also limits the ability to control invasive plant infestations and restore native plant communities. Terrain can preclude the use of conventional restoration tools and often results in the use of more costly and/or less successful methods. Complete restoration of some wildlands may also be limited by a lack of propagule sources for all the native plant species displaced by the invader. Large amounts of resources have been intensively spent on efforts to restore a few infested wildlands, yet invasive plants continue to spread as emerging populations and expanding established infestations. The conventional reactive approach to invasive plant species management has been largely ineffective at reducing impacts of invasive plants on wildlands, suggesting there is a critical need to adopt a different strategy. A comprehensive strategy with a primary focus of proactively preventing invasive plant establishment and spread will be required to successfully reduce the impacts of invasive plant species on wildlands.

Figure 1. *Taeniatherum caput-medusae* (annual grass dominating the community) invaded *Artemisia tridentata*-bunchgrass plant community (left) and non-invaded *Artemisia tridentata*-bunchgrass plant community (right). Note the lack of diversity in plant species composition and structure in invaded community compared to the non-invaded plant community. Also note the greater fine fuel continuity and quantity in the invaded compared to non-invaded plant community. This fuel loading can result in an increased fire frequency that is determinantal to native vegetation (*Artemisia tridentata* skeleton at top of invaded community).

Adopting an approach to managing invasive plant species that focuses on preventing new infestations and limiting expansion of existing infestations has the potential to preclude the need for restoration on millions of wildland hectares. However, this does not eliminate the need to restore already invaded wildland plant communities, but suggests that with current resource and knowledge constraints that prevention is a more logical priority than restoration in most circumstances. Preventing invasive plant infestations would be more cost-effective and successful than restoration attempts after invasive plant have established infestations (Peterson and Vieglasis 2001; Simberloff 2003; Zavaleta 2000). In fact, the Office of Technology Assessment (1993) estimated that every dollar spent on prevention and early control of invasive plants prevented $17 in later expenses. Prevention is an essential component to developing a successful invasive plant management programs (Sheley et al. 1996; DiTomaso 2000; Davies and Sheley 2007a). Furthermore, the cumulative benefits of preventing the establishment of invasive plant species infestations are exponential. Each infestation prevented eliminates all the future descendant infestations.

However, prevention is probably the most under-utilized strategy for managing invasive plant species in wildlands. The limited deployment of prevention may be a product of the lack of information detailing how to develop successful prevention programs. Most prevention discussions, excluding Davies and Sheley (2007a), are little more than a list of common-sense considerations (e.g. Sheley et al. 1999; Westbrooks et al. 1997). We propose that successful prevention of invasive plant infestations should focus on three primary strategies: 1) limiting the spatial dispersal of propagules (i.e. reducing propagule pressure), 2) maintaining or increasing the ability of wildland plant communities to resist invasion (i.e. biotic resistance), and 3) systematically searching for and eradicating emerging infestations.

Limiting Spatial Dispersal

Invasive plant propagules (most often seeds) create new infestations by being spatially dispersed to new areas. Limiting the spatial dispersal of invasive plants can greatly improve prevention success (Davies and Sheley 2007a). If invasive plants do not have propagules present at a site or are unable to disperse propagules to the site, invasion will not occur. Furthermore, the likelihood of invasion is significantly less with decreasing propagule pressure (D'Antonio et al. 2001; Davies 2008; Davies et al. 2008).

National and regional regulations to prevent the introduction of exotic species play an important role in limiting the spatial dispersal of invasive plants into new geographic regions. However, once an invasive species establishes in a geographic region, management becomes the primary means to preventing its spread within that region. Thus, our discussion is focused on research and management to prevent the spread of regionally-established invasive plant species.

One strategy to limit the spatial dispersal of invasive plant species is to reduce the amount of propagules they produce. Depending on the invasive plant species and location of infestations, various methods can be deployed to accomplish this task. Grazing or mowing prior to propagule development, introduction of biological control agents, and decreasing resource availability can potentially limit spatial dispersal by reducing propagule production. Defoliation by mowing or grazing of invasive plant species prior to propagule development causes a shift in resource allocation away from reproduction to growth (DiTomaso 2000; Sheley et al. 1998). The closer defoliation occurs to the end of the growing season, the less likely the invasive plant will have enough time and resources to grow new foliage and reproduce. Biological control agents can also be a defoliator that reduces the general production of the invader or they can be propagule predators. For example, two biological control insects reduced *Centaurea solstitialis* L. (yellow star-thistle) seed production by 43-76% (Pitcairn and DiTomaso 2000). A potential method to reduce resource availability to invasive plants is to practice management that increases competition for resources by other plants growing in the infestations. These tools to reduce propagule production of invasive plant species are only efficient under specific circumstances and often other management actions are needed.

Additional general management, such as applying herbicides around the edge of established invasive plant infestations can limit their expansion (Sheley et al. 1999) or defoliating invasive plants at the edge of infestations prior to seed development can reduce the spatial dispersal of invasive plant species without expending resources to control the entire infestation. However, specific management focused on restricting dispersal mechanisms can further improve the effectiveness of a prevention program. Limiting the spatial dispersal of invasive plant species can be accomplished by identifying and developing management that obstructs dispersal vectors. Most invasive species have special adaptations that facilitate their spatial dispersal by specific vectors. Davies and Sheley (2007a) identified seed characteristics (potential adaptations for dispersal) and location of the infestations relative to vector pathways as critical elements to consider in determining which vectors are major contributors to the spread of the invasive plant species (Fig. 2). Specific strategies could then be focused on these vectors to improve the effectiveness and efficiency of prevention efforts. Factors that promote specific dispersal are attachment, attractant, aerodynamic properties, buoyancy, and self-propelled adaptations, and infestation location (Davies and Sheley 2007a).

Invasive plants may have adaptations that facilitate attachment to vectors to increase spatial dispersal. Plants that are dispersed by animals and humans often have barbs, hooks, or viscid outgrowths on their seeds (Sorensen 1986; Cousens and Mortimer 1995). These outgrowths allow them to attach to clothing, hair, and fur and thus, can be dispersed great distances (Cousens and Mortimer 1995). Some invasive plants are dispersed by attracting potential vectors. Seeds that are edible or surrounded by edible flesh would likely be dispersed by animals and humans. These seeds may be consumed and then dispersed in animal feces. For example, horses can be major disperses of invasive plants species by transporting seeds in their digestive tract (Wells and Lauenroth 2007). Seeds surrounded by edible flesh may also be transported by humans and animals to a new location for consumption, the edible flesh is consumed, and then the remaining seed is discarded. Limiting dispersal of invasive plants that are commonly transported by animals and humans should focus on limiting their contact with potential vectors when propagules can be dispersed. Physical barriers to animal and humans can decrease their contact with the invasive plant species and thus reduce the spread of these plants. Feeding animals invasive plant-free feed in containment prior to moving them from locations with infestations to non-infested areas would also reduce the spread of invasive plants.

Many invasive plant species have seeds with aerodynamic properties that enhance their ability to be dispersed by wind (Fig. 3). Wind dispersed seeds have plumes or wing appendages (Burrows 1986); alternatively, whole plants or their panicles may disarticulate before seed shatter and be wind dispersed. For example, entire above ground structure of tumbleweeds breaks off and can be dispersed by wind (Howe and Smallwood 1982, Cousens and Mortimer 1995). Invasive plants may also have mechanisms for self-propelled dispersal. *Euphorbia esula* L. (leafy spurge) self-propels its seeds away from the parent plants with capsules that erupt during periods of low humidity and hot temperatures (Selleck et al. 1962). Self-propelled seeds may be projected several meters (Riley 1930; Beattie and Lyons 1975).

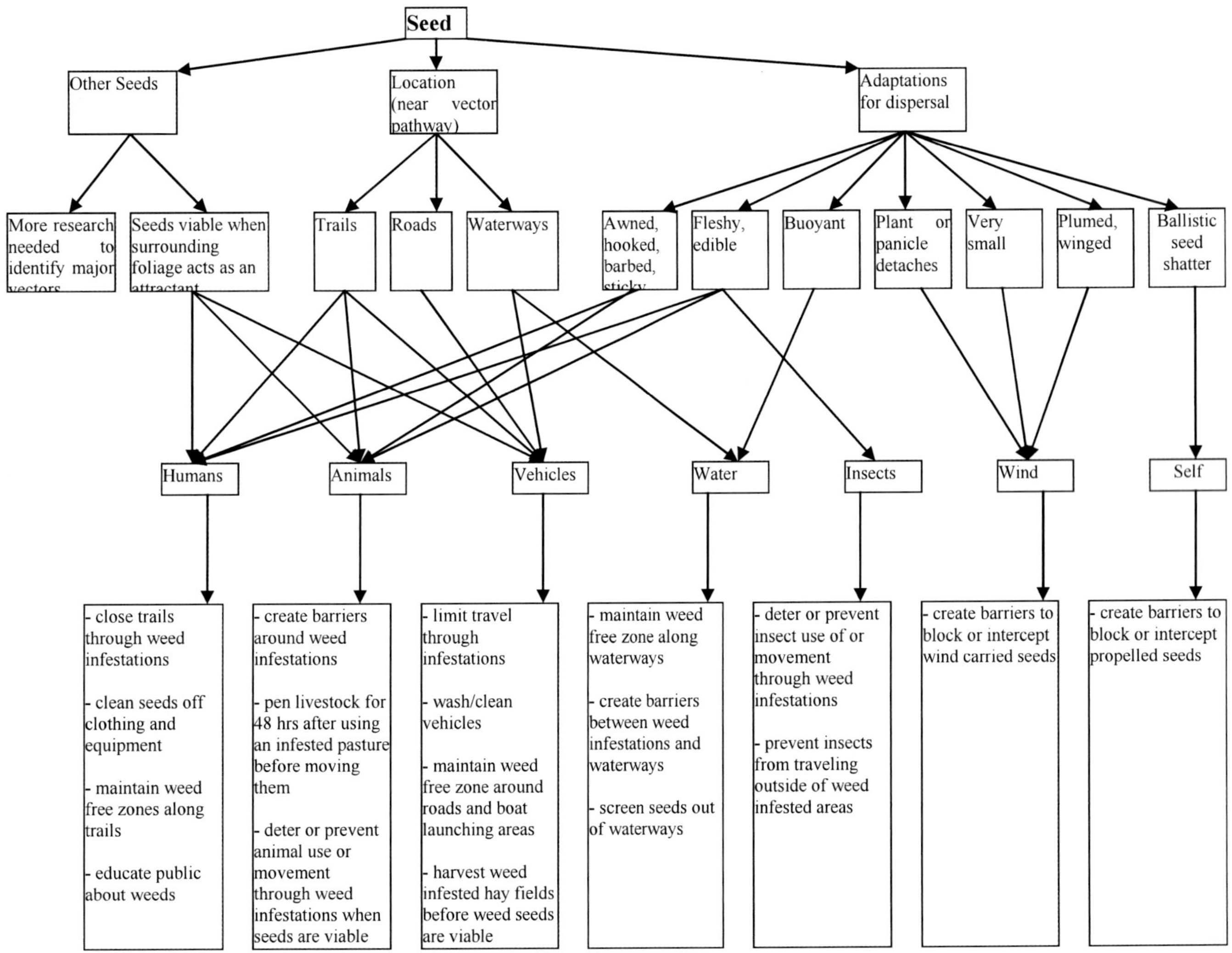

Figure 2. A framework for identifying potential major spatial dispersal vectors of an invasive plant species and management strategies to limit specific dispersal. (Copyright © 2007 Weed Science Society of America: Reprinted from Davies, K.W., and R.L. Sheley. 2007. A conceptual framework for preventing the spatial dispersal of invasive plants. Weed Science 55:178-184 by permission of Allen Press Publishing Services).

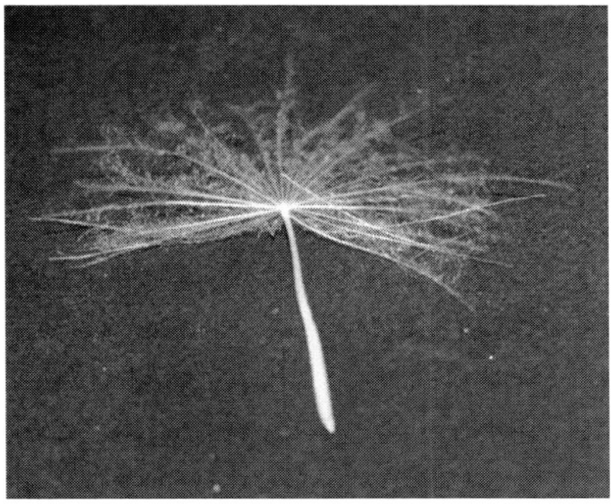

Figure 3. A seed with aerodynamic properties that are an adaptation for wind-dispersal. Note the high surface area of the plumes compared to the seed size.

Invasive plants with adaptations for wind- and self-propelled dispersal can be managed with similar strategies. The focus should be to intercept the aerial projected seeds or structures containing seeds. Davies and Sheley (2007b) demonstrated that wind-dispersal of invasive plants could be significantly reduced by maintaining taller neighboring vegetation around invasive plants.

Buoyant invasive plant propagules have the potential to be water dispersed. Some seeds can float for months (Hocking and Liddle 1986; Schneider and Sharitz 1988) increasing their dispersal potential and becoming a greater challenge to management. Small seeds can also be water dispersed because the amount of energy required to transport them is small. Strategies for limiting water-dispersal of invasive plants should strive to keep the propagules from reaching the water, removing them from the water, and/or stopping them from traveling from the water to a safe site while still viable (Davies and Sheley 2007a).

Locations of invasive plant infestations relative to potential vector pathways are also important factors to consider when managing invasive plant species. Infestations near roads and animal trails can accentuate dispersal by vehicles (Plummer and Keever 1963) and animals (Darwin 1859) even if plants lack specific adaptations for those methods of dispersal. To prevent dispersal along vector pathways, management needs to limit contact between vectors and invasive plants when dispersal can occur. Intensive management of infestations along roads, trails, and waterways to create an invasive plant-free zone next to the vector pathway would be justified. Closing or rerouting remote hiking trails and dirt roads that intersect invasive plant infestations may also be necessary.

Another major factor contributing to the spread of invasive plants is the dispersal of their propagules with desirable plant materials. Dispersal of invasive plant propagules with hay, other feeds, and seed mixes is a serious management concern. Seeds of *Centaurea* L. (knapweeds) species are often dispersed with hay and seed mixes (Sheley et al. 1998). Thus, recreational activities that are seemingly innocuous, such as using hay or grain on backcountry horse-packing trips could easily spread invasive plants to remote wildlands. Requirements for certified weed-free hay, other feeds, and seed on wildlands are probably the only effective strategy for curtailing this threat.

To limit the spatial dispersal of invasive plant species propagules, combinations of strategies will probably need to be applied because many invaders are dispersed by more than one mechanism or more than one invader is of concern. For a more detailed description of spatial dispersal of invasive plant species and suggested management options refer to Davies and Sheley (2007a). Preventing the spatial dispersal of invasive plants species in wildlands will also require educating the general public and, specifically, users of wildlands about invasive plants and implementing regulations to prevent invasive species introduction. Limiting the spatial dispersal of invasive plant propagules could be improved with additional research developing new methods and strategies to reduce propagule production, intercept dispersing propagules, and prevent vector contact with invasive plants.

Plant Community Resistance

The spread of invasive plant species across wildlands is dependent not only on the spatial dispersal of its propagules, but also whether or not these propagules can become established in existing plant communities. Discussions of prevention or any other approaches to invasive plant management are of little consequence if the invasive plant species is dispersing to an environment unsuitable for its survival. Thus, the present discussion focuses on environments that the invasive plant species can survive and thereby invade. The ability of a plant community to limit establishment of invasive plant species can be thought of as "plant community invasibility" or "biotic resistance to invasion". Biotic resistance is influenced by species diversity, functional group diversity, dominance, and disturbance. The contribution of these factors to biotic resistance is their influence on the resource sequestering in relation to resource availability of the existing plant community. Decreasing biotic resistance increases safe sites and other resources that are available for invasive plant species to capitalize on and become established.

Because negative relationships between native and invasive plant diversity have been documented (Elton 1958; Tilman 1997; Knops et al. 1999; Levine 2000; Brown and Peet 2003), diverse plant communities have been assumed to be more resistant to invasion than less diverse plant communities. However, at large spatial scales more diverse plant communities tend to have less resistance to invasion (Lonsdale 1999; Stohlgren et al. 1999, 2003). This discrepancy in the impact of diversity on plant community invasibility may be the result of spatial heterogeneity in the environment driving species diversity in landscapes, thus landscapes with more environmental diversity can support a greater diversity of native *and* exotic plant species (Davies et al. 2005). Similarly, temporal heterogeneity in climate conditions could increase the opportunities for the establishment of a greater diversity of plant species.

Plant functional group diversity may be a better indicator of plant community resistance to invasion than species diversity. Plant functional group diversity has profound implications to community stability (Hooper and Vitousek 1997; Tilman et al. 1997; Hooper 1998; Dukes 2001; Davies et al. 2007). Davies et al. (2007) and Pokorny et al. (2005) suggested that functional group diversity is critical to plant community resistance to invasion. Plant communities with higher diversity of functional groups used more resources than plant communities with fewer functional groups (Davies et al. 2007).

However, some plant communities may have only one dominant functional group that is important to reducing invasibility (Davies 2008). Thus, dominant plant species may also be important in determining plant community resistance to invasion. In a tallgrass prairie plant community, dominance was of greater importance to determining invasibility than species richness (Smith et al. 2004). Jiang et al. (2007) also reported that species composition was important to determining invasibility. Further supporting this assertion is that disturbances that damaged dominant species increased survival of invaders compared to disturbances that did not damage dominants (Burke and Grime 1996).

Invasibility of plant communities is also influenced by disturbance. Resistance to invasion decreased as disturbances increased bare ground (Burke and Grime 1996). Disturbances have been generally assumed to increase resource availability to invaders and disturbance prevention has been suggested to reduce the potential for exotic plant invasion (Sheley et al. 1999, Clark 2003). However, disturbances may be needed in some plant communities to maintain their resistance to invasion (Davies et al. 2008). Preventing disturbances can result in the loss of important mechanisms that allow plant communities to adapt to external pressure (Groffman et al. 2006).

Invasibility is clearly not controlled by a single casual factor, but by interactions among them. For example, community structure (Smith and Knapp 1999) and composition (Burke and Grime 1996) interact with disturbance to influence plant community invasibility. Relationships between plant community characteristics and invasibility likely differ among plant communities (Jiang et al. 2007). Plant communities that are resistant to invasion are not defined by one measurable characteristic. However, invasion resistant plant communities need to use resources both spatially and temporally to preclude their use by invaders. Resource use can be enhanced by limiting catastrophic disturbances, maintaining appropriate disturbance regimes, and preserving dominant species and major functional groups. Catastrophic disturbances produce opportunities for invasive plant species to establish by removing vegetation (particularly dominant species or major functional groups) that limit the availability of resources to invasive plants. Similarly, the disturbance regime can greatly influence plant community dynamics. Plant communities may be tolerant of periodic disturbances, but intolerant of the same disturbance at a higher frequency of occurrence. Interactions between disturbances can also be critical to determining the response of the plant community (Miller 1982; Collins 1987). Thus, the response to the combination of disturbances and their frequency (i.e. the current disturbance regime) the plant community will be subjected to may be more important than its response to any single disturbance. The dominant species or major functional groups use the most resources, thus, losing them from the plant community can result in a flush of resources for invasive plant species to capitalize on. Therefore, the ability of the dominant plant species or major functional groups to tolerate the current disturbance regime is of critical importance for maintaining biotic resistance to invasion. Management should focus on methods for improving tolerance of the plant community to the existing disturbance regime or altering the disturbance regime to favor survival of dominant plant species and major plant functional groups.

Searching for and Eradicating

Strategically searching for and eradicating new infestations is a critical component of a successful prevention program because uncontrollable and random events may still lead to new infestations even with the best management of community biotic resistance and invasive plant dispersal. Searching for and controlling new invasive plant infestations is a more effective strategy than trying to control large infestations (Moody and Mack 1988; Smith et al. 1999). However, time and resource limitations coupled with the vast nature of wildlands make searching all areas for new infestations untenable. This suggests a critical need to employ a strategic approach for planned searches. Areas where invasions are likely to occur should receive priority for planned searches. One of the most logical ways to systematically search for new infestations is to search around established infestations and along potential vector pathways. For example, horses using recreation trails can be major long-distance dispersal agents of invasive plants (Wells and Lauenroth 2007), suggesting recreation trails are a potential vector pathway and a logical choice for planned searches. If an invasive species is primarily wind dispersed, searching downwind of established infestations would provide the most return on resources expended. Searching areas that have been heavily disturbed should also receive priority because only a few propagules are required to establish a new invasive plant infestation when biotic resistance is low (D'Antonio et al. 2001).

An economical tool for increasing detection of new infestations is to capitalize on opportunities where inventory of new infestations can occur as a secondary activity with other planned activities. For example, in rented grazing allotments, livestock producers could be educated about existing invasive species infestations and/or new invasive plant threats. Livestock producers and/or their employees could then document any new infestations located during normal livestock management activities. When livestock are removed from the grazing allotment, livestock producers would be contacted to obtain any information gathered. This is just one example of several existing opportunities to search for new infestations across large areas while expending few resources. Other opportunities include using recreationists, wildland fire fighters, university students, and road maintenance personal to inventory invasive plant infestations during their normal activities. This highlights the need for public education and awareness of invasive plants.

Once new infestations are located, eradication should be attempted. The smaller the infestation and earlier it is detected, the greater the chance for successful eradication. Eradication efforts are usually confined to infestations smaller than one hectare (DiTomaso, 2000); however, larger scale eradication projects are possible. Larger scale eradication projects are usually attempted on invasive species that have relatively few populations but threaten economic and ecological losses across vast areas. The goal of eradicating new invasive plant infestations is to limit reproduction and subsequent development of a seed bank. Intensively managing small infestations for eradication can protect much larger areas (Pokorny and Krueger-Mangold 2007). Eradication should be viewed as an iterative approach in which the original treatments are monitored for several years and retreated if necessary. Follow-up monitoring, even if the eradication effort appears successful, will be required to ensure that invasive plants do not return from the seed bank and to prevent other invaders from occupying the spaces vacated by the eradicated invasive plant and/or non-target plants that suffered mortality from the eradication treatment.

Implementing Invasive Plant Prevention

Successfully preventing the spread of invasive plants on wildlands requires understanding what factors are allowing invasive plants to be successful. The success of invasive plants depends on their ability to disperse enough propagules to overcome the biotic resistance of the plant community they are trying to invade (D'Antonio et al. 2001; Davies 2008; Davies et al. 2008). With high biotic resistance the invader's propagule pressure has to be exceedingly large for successful invasion; however, at low biotic resistance even a few propagules can produce a successful invasion (D'Antonio et al. 2001). Thus, the first step in preventing invasion is to determine the biotic resistance of the plant community and propagule pressure from the invasive plant species. Determining the relationship that exists between propagule pressure and biotic resistance allows for an assessment of where prevention efforts need to be focused (Fig. 4). For example, if the plant community is fairly resistant to invasion, but propagule pressure is large from a nearby infestation, the most prudent decision would be to focus management actions on reducing propagule dispersal and early detection/eradication. It would be ineffective to expend resources and efforts to make small gains in community biotic resistance. However, if the plant community has little resistance to invasion and low invasive plant propagule pressure, then it is imperative to implement management that increases the communities' resistance to invasion and to eradicate new infestations. If the resistance of a plant community cannot be appreciably improved, then a comprehensive program to limit the introduction of invasive plants and to locate and eradicate new infestations must be implemented. The ecology of the invasive plant species and the risk of the plant communities to invasion must be considered to decide on the most appropriate course of action.

Propagule Pressure	Biotic Resistance: Low	Biotic Resistance: High
High	**High Invasion Risk** -All three strategies should be employed	**Moderate Invasion Risk** -Limit spatial dispersal of invader -Search for & eradicate new infestations
Low	**Moderate Invasion Risk** -Increase plant community resistance -Search for & eradicate new infestations	**Low Invasion Risk** -Search for & eradicate new infestations - Use of other methods are likely inefficient

Figure 4. A guide to deciding which management strategies would be most effective at preventing invasive plant species infestations and where the risk of invasion is greatest based upon the biotic resistance of the existing plant community and propagule pressure of invasive plant species.

Successful implementation of invasive plant prevention programs also requires developing wide spread support among land managers and land users. Invasive plant infestations are not impeded by boundaries in land ownership, thus the most successful prevention programs will also not be limited by property boundaries or ownership categories. Cooperative weed management areas (CWMA) may be effective levels to implement multiple-landscape prevention programs, while more specific management may be refined at the watershed level. Prevention can be implemented by adopting adaptive management strategies, where invasive plant prevention across landscapes can be modified based on monitoring the success of previous management actions.

Need for Additional Research

Development of more effective tools for limiting the dispersal of invasive plants could greatly improve prevention efforts (Davies and Sheley 2007a). This includes further identification of major vectors and methods for reducing their effectiveness at spreading invasive plants. Research could also improve dispersal prevention by determining how effective different strategies and combinations of strategies are at reducing spatial dispersal of invasive plants.

Identifying more characteristics and interactions among characteristics of plant communities that are resistant to invasion is needed to provide more concrete management objectives. Another important characteristic that may be a potential contributor to invasion is temporal climatic variation and its interaction with biotic resistance. Research investigating the influence of climatic variations on invasibility may provide valuable insights into assessing invasion risk. Development of a few measurable attributes that could be used to estimate plant community biotic resistance to invasion would be invaluable. This would allow for rapid, accurate predictions of invasion risk and for prioritization of management actions according to invasibility. Linking this information with geo-spatial technology could produce maps for planning at large spatial scales.

Advancements in remote sensing are encouraging, but more refinement is needed before the technology is ready for reliable detection of new invasive plant infestations. Currently, remote sensing is more applicable for identifying large scale established infestations. Improving models to predict invasive plant spread, especially across large landscapes, will improve early detection of new infestations and allow for more strategically implemented prevention strategies.

Research investigating how to manage plant communities to promote post-disturbance resistance to invasive plant invasion would also be invaluable. The interactions among disturbances could have substantial effects on the biotic resistance of the plant community to invasion. Interactions among disturbances are important determinants of plant community characteristics (Collins 1987). Thus, pre-disturbance management could have significant implications to the biotic resistance of the plant community after disturbance and warrants investigation.

Conclusion

Active prevention programs are essential to effective invasive plant management. Invasive plant prevention programs can protect plant communities that either cannot be restored or are prohibitively expensive to restore if invasive plants dominate. Prevention is also more cost effective than trying to control already invaded plant communities, because prevention precludes the need for highly expensive and often unsuccessful management practices. Limiting the spatial dispersal of invasive plants and increasing or maintaining the resistance of the plant community to invasion can significantly decrease the probability of new infestations. However, systematically searching for and eradicating new infestations is an important component of any successful prevention program, because complete dispersal prevention is unlikely and random events may allow opportunities for invasive plants to establish in relatively invasion resistant plant communities. Additional research can improve prevention success by providing better tools and strategies. Prevention needs to receive greater priority in invasive plant management and research.

REFERENCES

Beattie, A.J., and N. Lyons. 1975. Seed dispersal in *Viola* (Violaceae): adaptations and strategies. *American Journal of Botany* 62:714-722.

Brown, R.L., and R.K. Peet. 2003. Diversity and invasibility of southern Appalachian plant communities. *Ecology* 84:32-39.

Burke, M.J.W., and J.P. Grimes. 1996. An experimental study of plant community invasibility. *Ecology* 77:776-790.

Burrows, F.M. 1986. The aerial motion of seeds, fruits, spores and pollen. pp. 1-47. *In:* Murray, D.R. (ed). *Seed dispersal*. Academic Press, Inc. Orlando, FL.

Clark, J. 2003. Invasive plant prevention guidelines. Center for Invasive Plant Management. Bozeman, MT. 15 p. www.weedcenter.org. Last accessed on 30 January 2006.

Collins, S.L. 1987. Interactions of disturbances in tallgrass prairie: a field experiment. *Ecology* 68:1243-1250.

Cousens, R., and M. Mortimer. 1995. *Dynamics of weed populations*. Cambridge University Press. New York, NY. 332 p.

Cronk, C.B., and J.L. Fuller 1995. *Plant invaders: the threat to natural ecosystems*. Chapman and Hall, New York, NY. 241 p.

D'Antonio, C., J. Levine, and M. Thomsen. 2001. Ecosystem resistance to invasion and the role of propagule supply: a California perspective. *Journal of Mediterranean Ecology* 2:233-245.

D'Antonio, C., and L.A. Meyerson. 2002. Exotic plant species as problems and solutions in ecological restoration: a synthesis. *Restoration Ecology* 10:703-713.

Darwin, C. 1859. The origin of species. John Murray, London, UK. 380 p.

Davies, K.F., P. Chesson, S. Harrison, B. D. Inouye, B.A. Melbourne, and K.J. Rice. 2005. Spatial heterogeneity explains the scale dependence of the native-exotic diversity relationship. *Ecology* 86:1602-1610.

Davies, K.W. 2008. Medusahead dispersal and establishment in sagebrush steppe plant communities. *Rangeland Ecology and Management* 61:110-115.

Davies, K.W., M.L. Pokorny, R.L. Sheley, and J.J. James. 2007. Influence of plant functional group removal on soil inorganic nitrogen concentrations in native grasslands. *Rangeland Ecology and Management* 60:304-310.

Davies, K.W., and T.J. Svejcar. (IN PRESS) Comparison of medusahead invaded and non-invaded Wyoming big sagebrush steppe in southeastern Oregon. *Rangeland Ecology and Management*

Davies, K.W., and R.L. Sheley. 2007a. A conceptual framework for preventing the spatial dispersal of invasive plants. *Weed Science* 55:178-184.

Davies, K.W., and R.L. Sheley. 2007b. Influence of neighboring vegetation height on seed dispersal: implications for invasive plant management. *Weed Science* 55:626-630.

Davies, K.W., R.L. Sheley, and J.D. Bates. 2008. Does prescribed fall burning *Artemisia tridentata* steppe promote invasion or resistance to invasion after a recovery period? *Journal of Arid Environments* 72:1076-1085.

DiTomaso, J.M. 2000. Invasive weeds in rangelands: species, impacts, and management. *Weed Science* 48:255-265.

Dukes, J.S. 2001. Biodiversity and invasibility in grassland microcosms. *Oecologia* 126:563-568.

Eiswerth, M.E., and W.S. Johnson. 2002. Managing nonindigenous invasive species: insights from dynamic analysis. *Environmental and Resource Economics* 23:319-342.

Elton, C.S. 1958. *The ecology of invasion by animals and plants*. Methuen, London, UK. 181 p.

Goebel, C.J., M. Tazl, and G.A. Harris. 1988. Secar bluebunch wheatgrass as a competitor to medusahead. *Journal of Range Management* 41:88-89.

Groffman, P.M., J.S. Baron, T. Blett, A.J. Gold, I. Goodman, L.H. Gunderson, B.M. Levinson, M.A. Palmer, H.W. Paerl, G.D. Peterson, N.L. Poff, D.W. Rejeski, J.F. Reynolds, M.G. Turner, K.C. Weathers, and J. Wiens. 2006. Ecological thresholds: the key to successful environmental management or an important concept with no practical application. *Ecosystems* 9:1-13.

Hironaka, M., and B.W. Sindelar. 1975. Growth characteristics of squirreltail seedlings in competition with medusahead. *Journal of Range Management* 28:283-285.

Hocking, P.J., and M.J. Liddle. 1986. The biology of Australian weeds: 15. *Xanthium occidentale* Bertol. Complex and *Xanthium spinosum* L. *Journal of the Australian Institute of Agricultural Science* 52:191-221.

Hooper, D.U. 1998. The role of complementarity and competition in ecosystem responses to variation in plant diversity. *Ecology* 79:704-719.

Hooper, D.U., and P.M. Vitousek. 1997. The effects of plant composition and diversity on ecosystem process. *Science* 277:1302-1305.

Howe, H.F., and J. Smallwood. 1982. Ecology of seed dispersal. *Annual Review of Ecology and Systematics* 13:201-228.

Huenneke, L. 1996. Ecological impacts of invasive plants in natural areas. Proceedings: *Western Society of Weed Science* 49:119-121.

Jiang, X.L., W.G. Zhang, and G. Wang. 2007. Biodiversity effects on biomass production and invasion resistance in annual verse perennial plant communities. *Biodiversity Conservation* 16:1983-1994.

Knops, J.M.H., D. Tilman, N.M. Haddad, S. Naeem, C.E. Mitchell, J. Haarstad, M.E. Ritchie, K.M. Howe, P.B. Reich, E. Siemann, and J. Groth. 1999. Effects of plant richness on invasion dynamics, disease outbreaks, insects abundances, and diversity. *Ecology Letters* 2:286-293.

Levine, J.M. 2000. Species diversity and biological invasions: relating local process to community pattern. *Science* 288:852-854.

Lonsdale, W.M. 1999. Global patterns of plant invasions and the concept of invasibility. *Ecology* 80:1522-1536.

Mack, R.N. 1981. Invasion of *Bromus tectorum* L. into western North America: an ecological chronicle. *Agro-Ecosystems* 7:145-165.

Mack, R.N., D. Simberloff, W.M. Lonsdale, H. Evans, M. Clout, and F. Bazzaz. 2000. Biotic invasions: causes, epidemiology, global consequences and control. *Issues in Ecology* 5:1-20.

Masters, R.A., and R.L. Sheley. 2001. Principles and practices for managing rangeland invasive plants. *Journal of Range Management* 54:502-517.

Miller, T.E. 1982. Community diversity and interactions between the size and frequency of disturbance. *The American Naturalist* 120:533-536.

Moody, M.E., and R.N. Mack. 1998. Controlling the spread of plant invasions: the importance of nascent foci. *Journal of Applied Ecology* 25:1009-1021.

Offices of Technology Assessment of United States Congress. 1993. Harmful non-indigenous species in the United States. US Government Printing Office, Washington, D.C. 391 p.

Peterson, A.T., and D.A. Vieglais. 2001. Predicting species invasions using ecological niche modeling: new approaches from bioinformatics attack a pressing problem. *BioScience* 51:363-371.

Pimentel, D., L. Lach, R. Zuniga, and D. Morrison. 2000. Environmental and economic costs of nonindigenous species in the United States. *BioScience* 50:53-65.

Pimm, S., and M. Gilpin. 1989. Theoretical issues in conservation biology. pp. 287-305. *In:* Roughgarden, J., R. May, and S. Levey (eds). *Perspectives in ecological theory.* Princeton University Press, Princeton, NJ.

Pitcairn, M.J., and J.M. DiTomaso. 2000. Rangeland and uncultivated areas: integrating biological control agents and herbicides for star thistle control. pp. 65-72. *In:* Hoddle, M.S. (ed.). *California conference on biological control II.*

Plummer, G.L., and C. Keever. 1963. Autumnal daylight weather and camphor-weed dispersal in the Georgia piedmont region. *Botanical Gazette* 124:283-289.

Pokorny, M.L., and J.M. Krueger-Mangold. 2007. Evaluating Montana's dry woad (*Isatis tinctoria*) cooperative eradication project. *Weed Technology* 21:262-269.

Pokorny, M.L., R.L. Sheley, C.A. Zabinski, R.E. Engel, T.J. Svejcar, and J.J. Borkowski. 2005. Plant functional group diversity as a mechanism for invasion resistance. *Restoration Ecology* 13:448-459.

Randall, J. 1996. Weed control for the preservation of biological diversity. *Weed Technology* 10:370-383.

Riley, H.N. 1930. *The dispersal of plants throughout the world.* Ashford: Reeves, Kent UK. 744 p.

Schneider, R.L., and R.P. Sharitz. 1988. Hydrochory and regeneration in a bald cypress-water tupelo swamp forest. *Ecology* 69:1055-1063.

Selleck, G.W., R.T. Coupland, and C. Frankton. 1962. Leafy spurge in Saskatchewan. *Ecological Monographs* 32:1-29.

Sheley, R.L., J.S. Jacobs, and M.F. Carpinelli. 1998. Distribution, biology, and management of diffuse knapweed (*Centaurea diffusa*) and spotted knapweed (*Centaurea maculosa*). *Weed Technology* 12: 353-362.

Sheley, R.L., M. Manoukian, and G. Marks. 1999. Preventing noxious weed invasion. pp. 69-72. *In:* Sheley, R.L., and J.K. Petroff (eds). Biology and management of noxious rangeland weeds. Oregon State University Press, Corvallis, OR.

Sheley, R.L. T. J. Svejcar, and B.D. Maxwell. 1996. A theoretical framework for developing successional weed management strategies on rangeland. *Weed Technology* 10:766-773.

Simberloff, D. 2003. Eradication – preventing invasions at the outset. *Weed Science* 51:247-253.

Smith, H.A., W.S. Johnson, J.S. Shonkwiler, and S.R. Swanson. 1999. The implications of variable or constant expansion rates in invasive weed infestations. *Weed Science* 47:62-66.

Smith, M.D., and A.K. Knapp. 1999. Exotic plant species in a C_4-dominated grassland: invasibility, disturbance, and community structure. *Oecologia* 120:605-612.

Smith, M.D., J.C. Wilcox, T. Kelly, and A.K. Knapp. 2004. Dominance not richness determines invasibility of tallgrass prairie. *Oikos* 106:253-262.

Sorensen, A.E. 1986. Seed dispersal by adhesion. Annual Review of Ecology and Systematics 17:443-463.

Stewart, G., and A.C. Hull. 1949. Cheatgrass (*Bromus tectorum* L.) – an ecologic intruder in southern Idaho. *Ecology* 30:58-74.

Stohlgren, T.J., D.T. Barnett, and J. Kartesz. 2003. The rich get richer: patterns of plant invasions in the United States. *Frontiers in Ecology and the Environment* 1:11-14.

Stohlgren, T.J., D. Binkley, G.W. Chong, M.A. Kalkhan, L.D. Schell, K.A. Bull, Y. Otsuki, G. Newman, M Bashkin, and Y. Son. 1999. Exotic plant species invade hot spots of native plant diversity. *Ecological Monographs* 69:25-46.

Tilman, D. 1997. Community invasibility, recruitment limitation, and grassland biodiversity. *Ecology* 78:81-92.

Tilman, D., J. Knops, D. Wedin, P. Reich, M. Ritchie, and E. Siemann. 1997. The influence of functional diversity and composition on ecosystem processes. Science 277:1300-1302.

Torell, P.J., L.C. Erickson, and R.H. Haas. 1961. The medusahead problem in Idaho. *Weeds* 9:124-131.

Vitousek, P.M., C.M. D´Antonio, L.L. Loope, M. Rejmanek, and R. Westbrooks. 1997. Introduced species: a significant component of human-caused global change. *New Zealand Journal of Ecology* 21:1-16.

Wells, F.H., and W.K. Lauenroth. 2007. The potential for horses to disperse alien plants along recreational trails. *Rangeland Ecology and Management* 60:574-577.

Westbrooks, R. 1998. Invasive plants, changing the landscape of America: fact book. Washington, D.C.: Federal Interagency Committee for the Management of Noxious and Exotic Weeds. 109 p.

Westbrooks, R.G., L. Otteni, and R.E. Eplee. 1997. New strategies for weed prevention. Conference Proceedings: Exotic pests of eastern forests. Nashville, TN. 8-10 April 1997.

Whisenant, S.G. 1990. Changing fire frequencies on Idaho's Snake River Plains: ecological and management implications. p. 4-10. *In:* McArthur, E.D., E.M. Romney, S.D. Smith, P.T. Tueller, comps. *Cheatgrass invasion, shrub die-off, and other aspects of shrub biology and management*. Las Vegas, NV.

Wittenberg, R., and M.J. Cock. 2001. Invasive alien species: a toolkit of best prevention and management practices. Global Invasive Species Prgramme, CABI Publishing, Wallingford, Oxon, UK. 228 p.

Young, J.A. 1992. Ecology and management of medusahead (*Taeniatherum caput-medusae* ssp. *asperum* [SIMK.] Melderis). Great Basin Naturalist 52:245-252.

Young, J.A., and F.L. Allen. 1997. Cheatgrass and range science: 1930-1950. *Journal of Range Management* 50, 530-535.

Young, K. and J. Mangold. 2008. Medusahead (*Taeniatherum caput-medusae* ssp. *asperum*) outperforms squirreltail (*Elymus elymoides*) through interference and growth rate. *Invasive Plant Science and Management* 1:73-81.

Zavaleta, E. 2000. Valuing ecosystem services lost to Tamarix invasion in the United States. pp. 261-300. *In:* Mooney, H.A., and R.J. Hobbs (eds). *Invasive species in a changing world*. Island Press, Washington, D.C.

Zavaleta, E.S., R.J. Hobbs, and H.A. Mooney. 2001. Viewing invasive species removal in a whole-ecosystem context. *Trends in Ecology and Evolution* 16:454-459.

In: Invasive Species: Detection, Impact and Control
Editors: C.P. Wilcox and R.B. Turpin
ISBN 978-1-60692-252-1

Chapter 4

CHARACTERIZING FIELD-LEVEL HYPERSPECTRAL MEASUREMENTS FOR IDENTIFYING WETLAND INVASIVE PLANT SPECIES

Nathan M. Torbick[a,c], Brian L. Becker[b], Jiaguo Qi[a] and David P. Lusch[a]

[a] Department of Geography, Center for Global Change & Earth Observation, 1405 S. Harrison, Michigan State University, 48823, USA
torbick@msu.edu, qi@msu.edu, lusch@msu.edu

[b] Department of Geography, 292 Dow Science Complex, Central Michigan University, 48859, USA
becke1b@cmich.edu

[c] Applied Geosolutions, 403 Kent Place, Newmarket, New Hampshire, 03857, USA

ABSTRACT

Resource managers can benefit from improved methods for identifying invasive plant species. The utilization of hyperspectral remote sensing as a tool for species-level mapping has been increasing and techniques need to be explored for identifying species of interest. The overarching objective of this paper was to investigate three distinct processing methodologies (i.e., Derivatives, Continuum Removal, and Shape Filter) to explore their potential for delineating wetland invasive plant species within the spectral domain of typical airborne hyperspectral sensors. Field-level hyperspectral data (350-2500nm) were collected for twenty-two wetland plant species in a wetland located in the lower Muskegon River watershed in Michigan, USA. Generally, continuum removed spectra were more similar than raw reflectance for the invasive species of interest according to the Jeffries-Matusita distance measure. Second-derivative analysis showed that the wavelength locations of absorption and reflectance features were consistent for all species and emphasized the NIR region for separation. The shape-filter was useful as a method to identify invasive species and showed that useful wavelength regions can vary depending on the species of interest and approach utilized. Using the shape-filter, *Lythrum salicaria*, *Phragmites australis*, and *Typha latifolia* possessed maximum separation (distinguished from other species) at the red edge (700nm) and water absorption region (1350nm), the near-infrared down slope (1000 and 1100nm), and the visible/chlorophyll absorption region (500nm) and red edge (650nm), respectively.

Key words: invasive plant species, coastal wetlands, hyperspectral, derivatives, continuum removal, shape-filter

1. INTRODUCTION

Invasive species are one of the largest threats to wetlands biodiversity and ecosystem functioning. In the US, invasive species are estimated to cause $120 billion dollars per year in environmental damage and associated control costs: *Lythrum salicaria* (purple loosestrife) alone is estimated to cost $45 million per year as it spreads at a rate of 115,000 ha/yr across wetlands in the US [1]. With approximately 50% of wetlands destroyed or altered globally as a result of human activities [2], preserving and restoring the ecological integrity of the remaining wetlands has become a priority [3-5]. In order to meet future goals for aquatic ecosystem integrity, resource managers and decision makers need information and improved methods for identifying invasive plants and monitoring control measures.

Satellite remote sensing has been a useful tool in providing general information on wetlands types [6]; however, both spatial and spectral resolutions have limited the level of detail ultimately required for comprehensive wetland assessments. Recent advances in sensor technology and remote sensing science have promoted an interest in hyperspectral data for mapping wetlands at the species level [7-11]. Advanced spectroscopic systems possess capabilities to capture data at narrow spectral bandwidths on the order of three to ten nanometers, while contiguously covering large portions of the spectrum (e.g., 350-2500nm). This allows for small variations in plant/substrate absorptance and reflectance to be recorded. Incorporating such relatively high spectral detail makes it possible to explore species separability and precise process monitoring [10,12].

In the last few years several studies have utilized hyperspectral data for wetlands mapping. A primary goal in these investigations was to develop methods to utilize the increased level of data supplied via hyperspectral instruments. These studies can be grouped into methods to identify wavelengths of particular utility (processing techniques to extract and identify the most useful bands) and evaluating and improving classification techniques to map species of interest.

For example, Becker et al. [8] performed derivative analysis to identify unique points of inflection along spectra for wetland plants in a Great Lakes coastal wetland and identified eight bands as possessing the most utility for separation. The bands are located across the visible (VIS) and near-infrared (NIR) portions of the spectrum and are affiliated with domains that represent unique biophysical characteristics. The red-edge was highlighted as having particular strength in separation. Artigas and Yang [7] analyzed samples from a New Jersey coastal wetland and also identified the NIR region using a discrimination metric and derivative analysis. This research concluded that monotypic stands of *Phragmites* could be identified by using the unique NIR response. Schmidt and Skidmore [10] conducted Mann-Whitney *U*-testing on field-level reflectance data from coastal salt marshes in the Netherlands and found wavelengths in the NIR to possess high abilities to statistically differentiate species.

Developing and improving classifications has also been a focus of hyperspectral wetland remote sensing. For example, Becker et al. [9] examined the optimal spectral and spatial resolutions for mapping Great Lakes coastal wetlands. A series of experiments tested different bands, band combinations, and pixels sizes in order to identify the most advantageous configurations to accurately map coastal wetlands. The results showed that narrow, strategically located bands were necessary to achieve acceptable resiliency levels when trying to limit the number of bands in order to maintain small pixel sizes. Li et al. [13] and Rosso et al. [11] used Spectral Mixture Analysis (SMA) and Multiple End-member (MESMA) techniques on airborne data to map coastal marshes in central California and found spectral similarity and increasing landscape heterogeneity to cause misclassification. Underwood et al. [14] found minimum noise fraction to outperform band ratio and continuum removal processing techniques when classifying varying densities of nonnative coastal plants of concern. Recently, Artigas and Yang [15] found that mapping *Phragmites australis* gradients in coastal New Jersey using airborne imagery was possible due to spectral differences associated with physiognomic characteristics. All these studies show the importance of utilizing unique spectral features to better differentiate/map wetland plant canopies.

A processing technique that allows for the extraction and modeling of individual spectral features (absorptance and reflectance) is continuum removal [13]. This technique is increasingly being implemented in hyperspectral vegetation investigations to isolate features of utility [10, 11, 17-19]. Used extensively in geological applications, continuum removal disregards albedo, and/or contributing background signal, to obtain individual features (absorption/reflectance) such as the precise location and depth of absorption features. By using continuum removal, Underwood et al. [14] achieved accurate maps of coastal invasives by applying continuum removal techniques to take advantage of unique water absorption features. Schmidt and Skidmore [10] found that applying continuum removal to salt-marsh vegetation spectra improved species separation in the visible spectrum, but decreased it in the near-infrared (NIR) and shortwave-infrared (SWIR) regions. They further suggest that if continuum removal eliminates noise from the soil background, moisture content, and canopy structure, then only the varying biogeochemical content of a species would determine separability levels. In this chapter, we explored the utility of continuum removal for identifying wetland invasive plant species.

One second well-established technique for characterizing spectra is derivative analysis. The methodology distinguishes the wavelength location where substantial inflection occurs [20-22]. In Becker et al. [8], second derivative approximations identified seven wavelengths (685, 731, 939, 514, 812, 835, 823, 560 nm) using contiguous data covering the visible and NIR regions from a Great Lakes coastal wetland community. In this chapter, we expanded upon the techniques applied in Becker et al. [8] to evaluate if invasives species of interest and different biological communities reproduce similar results.

The third identification approach utilized vegetation reflectance variation for distinguishing wetland invasive plant species. This approach, developed by Cochrane [23], uses a *shape-filter* representing the range of reflectance variation present in a species over the spectrum. Vegetation reflectance varies across the spectrum with the visible domain largely determined by the chlorophyll content, the NIR region a function of leaf structure and biomass volume, and the SWIR region largely determined by leaf water content and biomass volumes [23, 24]. Generally, differences in absorption determine the amount of variation and spectral overlap between species. The maximum and minimum reflectance for a given species creates its *shape-space.* If any of the comparison spectra (e.g., an invasive species vs. all others) fall outside the shape-space, then separation is possible. As absorption feature 'uniqueness' increases, the level of separability increases. As a byproduct, the shape-filtering technique identifies wavelengths that are useful for distinguishing between the species analyzed. The wavelengths identified might vary depending on species and uniqueness of absorption features. In this chapter we applied the shape filtering approach to identify invasives and wavelengths that were useful for doing so.

The overarching goal of this chapter was to investigate techniques for highlighting locations of unique spectral features for identifying wetland invasive plant species. This investigation utilizes three distinct processing methodologies (i.e., Derivatives, Continuum Removal, and Shape Filter) to explore their potential for delineating invasive species within the spectral domain of typical airborne and satellite hyperspectral sensors.

2. METHODS

2.1. Study Area

The investigation was carried out in a wetland complex covering approximately 10 km^2 in the lower Muskegon River Watershed (MRW) located in west-central Lower Michigan, USA (W86° 09' 45", N43° 16' 10"). Fed from the Muskegon River and an extensive tributary network, this wetland complex serves as the last land-water interface before draining into Lake Muskegon, which then flows directly into Lake Michigan. Lake Muskegon (1,679 ha) was recognized as an Area Of Concern by the 1987 Great Lakes Water Quality Agreement due to declining aquatic ecosystem conditions. The National Wetland Inventory classifies the majority of the wetland complex as palustrine containing seasonally and semi-permanently flooded regions of scrub-shrub, forest, and emergent covers. Land uses adjacent to the wetland complex include residential and commercial neighborhoods, urban parks, industrial zones including a pulp and paper mill, chemical and petrochemical facilities, and agricultural and forest patches.

2.2. Data Collection and Preprocessing

A field campaign was conducted during mid-August (2006), which generally represents the peak of the growing season for wetland vegetation within the study area. Capturing data during peak phenological cycles has been shown to increase the separability of invasive wetland plants [25]. Due to the expansive nature of the wetland complex and the challenge of moving through a wetland, a compromise between operational feasibility and statistical sampling rigor was required. It was not the objective of this research to collect an extensive spectral library capturing the phenological and/or regional spectral differences displayed by common wetland plants. Rather, it was our intent to test the applicability of three processing techniques on a small but well controlled number of plant spectra. Both logistical constraints (equipment setup and takedown) and traveling throughout the wetland complex required a substantial amount of time. Reconnaissance field work identified emergent pools where high biodiversity and ecologically noteworthy species of interest (i.e., invasives) were present. Focusing our efforts around these regions of the wetland complex allowed a representative set of species spectra to be collected. An airboat provided the most efficient access for traveling around the wetland complex. Data acquisition focused on the dominant terrestrial-, emergent-, and submergent- species. Dominance was qualitatively identified during reconnaissance field work by evaluating percent cover and the approximate size of a patch for a species. A total of twenty-two wetland plant species were recorded (Table 1). Eight species are identified as being invasive [26]. Note that not all the species identified as invasive are classified as exotic and the degree of 'invasiveness' can vary by region and conditions.

Table 1. Plant species sampled in the study area

Scientific name	Common Name	Invasive
Sagittaria latifolia	Arrowhead, broadleaf	Invasive
Scirpus validus	Bulrush, softstem	
Typha latifolia	Cattail, broad-leaved	Invasive
Leersia oryzoides	Cutgrass	
Vallisneria americana	Eelgrass	
Eleocharis rostellata	Spikerush, beaked	
Elodea canadensis	Canadian waterweed	Invasive
Sparganium androcladum	Bur-reed, branched	
	Filamentous green algae	
Heteranthera dubia	Water star grass	
Iris versicolor	Iris, harlequin blue flag	
Lemna minor	Common duckweed	
Myriophyllum verticillatum	Watermilfoil, whorl-leaf	
	Mowed field grass	
Polygonum pensylvanicum	Pennsylvania smartweed	Invasive
Phragmites australis	Common reed	Invasive
Pontederia cordata	Pickerelweed	Invasive
Potamogeton spirillus	Spiral pondweed	
Lythrum salicaria	Purple loosestrife	Invasive
Nymphaea odorata	White water lily	Invasive
Salix eriocephala	Willow	
Nuphar lutea	Yellow pond lily	

We used a portable spectroradiometer (FieldSpec Pro FR®, Analytical Spectral Devices, Inc., Boulder, Colorado) to collect *in situ* radiance between 350-2500 nm (visible to shortwave infrared). Spectral resolution (full width half maximum) was approximately 3 nm in the visible wavelengths and 10 nm in the infrared region. The sensor was equipped with a 24 degree field-of-view (FOV) optic and held approximately 1-meter above the target at nadir for measurements representing field-canopy conditions. Sun-target-sensor geometry was repeated as best as possible under these difficult field conditions between 11:00-14:30 local time. This viewing geometry configuration approximately represents the spatial resolution that current airborne hyperspectral sensors can achieve (~approximately 1m). A Spectralon® panel (Labsphere, Inc., North Sutton, New Hampshire) was used for calibration during processing and atmospheric adjustments. Eight rapidly sequenced measurements were averaged for one spectrum and eight or nine spectra (depending on the amount of non-overlapping FOVs within a patch) were collected in a homogeneous (one species) plot. The instrument was shifted within the patch during collection to capture inherent within patch variability and ensure non-overlapping FOVs. This was repeated for four patches for each plant. This process provided 32 unique measurements for each species. During data acquisition, the sensor was first placed over the reference panel to record the panel-reflected radiance. Then the sensor was placed over the target to record the target-reflected radiance. Then, by ratioing the radiance measurements, the surface *reflectance factor* was calculated. By definition, the term *reflectance factor* is the ratio of radiant emittance of a target (i.e., wetland plant) to that reflected into the same reflected-beam geometry and wavelength range by an ideal and diffuse standard surface (i.e., the Spectralon panel) irradiated under the same conditions [27]. The reflectance factor was calculated based on the following equation:

$$\rho = tE\uparrow / cpE\uparrow \tag{1}$$

where ρ is *in situ* reflectance factor for target of interest (wetland plant species), $tE\uparrow$ is target (wetland plant species) *in situ* radiance, and $cpE\uparrow$ is the calibration panel *in situ* radiance.

Subsequent data processing in this study also removed wavelength regions severely affected by atmospheric absorption in the spectral ranges of 1350-1480, 1775-2000, and >2400 nm [22]. Figure 1 displays invasive spectra.

2.3. Analytical Techniques

A variety of measures of separation exist that quantify the degree of dissimilarity between any two probability distributions. One commonly applied separation measure is the Jeffries-Matusita (JM) distance measure. A few versions of the JM distance equation exist throughout the remote sensing literature [10, 28]. For the purposes of this analysis, the JM distance equation presented by Niel et al. [28] was utilized. In this version, a JM distance value can range from 0 (i.e., identical distributions) to 1.414 (i.e., complete dissimilarity). We used the JM distance measure (Eq. 2 & 3) to evaluate spectral separability and assess the process of continuum removal. :

$$JMij = \sqrt{2(1-e^{-a})} \tag{2}$$

where $JMij$ is the Jeffries-Matusita distance between signatures i and j and a is the Bhattacharyya distance.

$$a = \frac{1}{8}(U_i - U_j)^T \left[\frac{C_i + C_j}{2}\right]^{-1} (U_i - U_j) + \frac{1}{2} Ln \left[\frac{1/2\left|C_i + C_j\right|}{\sqrt{|C_i| X |C_j|}}\right] \tag{3}$$

where i and j represent the two classes of interest, T is transpose, C_i is the variance-covariance matrix of signature i, U_i is the mean vector of signature i, $|C_i|$ is the determinant of C_i.

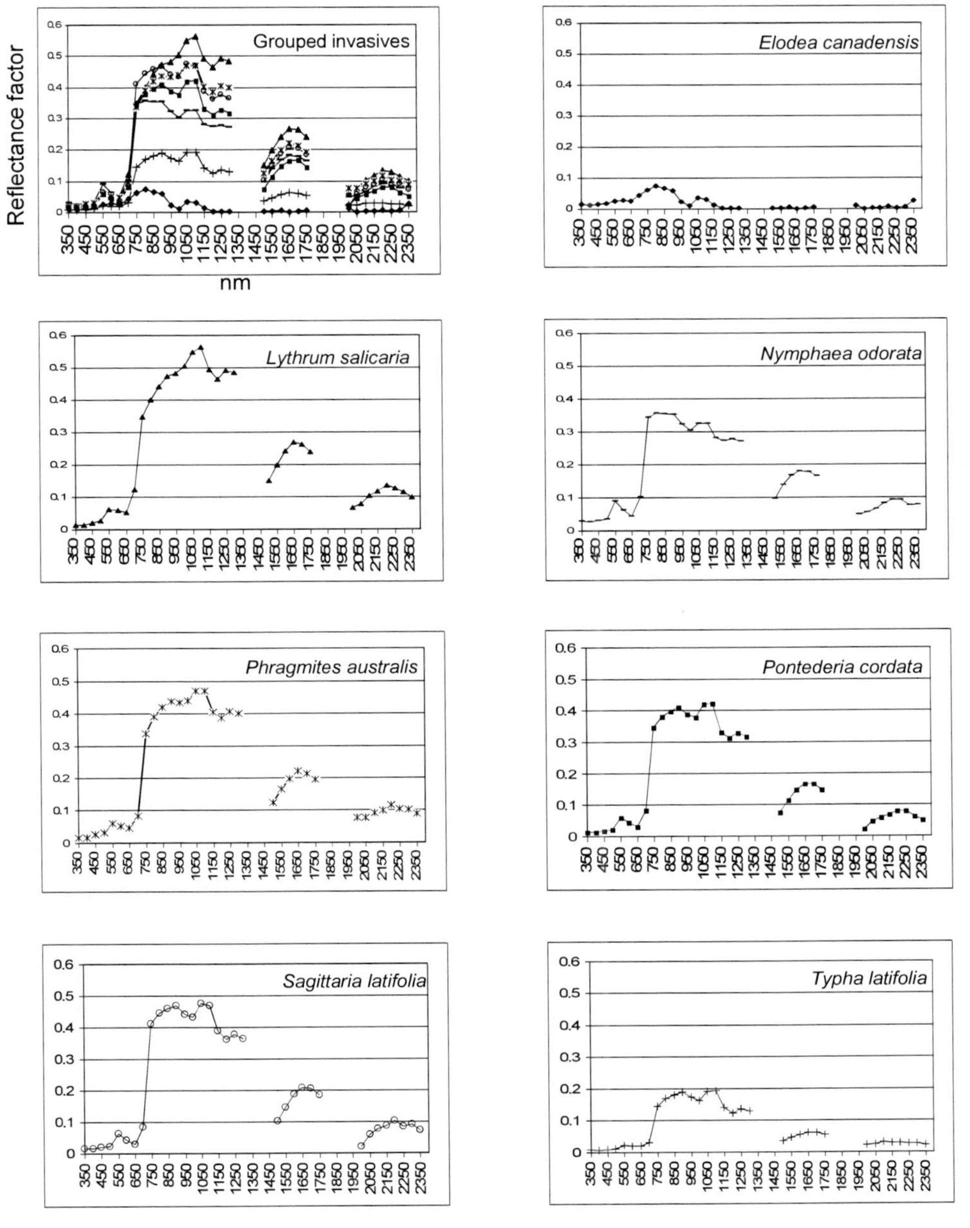

Figure 1. Reflectance factor at 50nm intervals for selected invasive plant species. Pre-processing removed wavelength regions severely affected by atmospheric absorption in the spectral ranges of 1350-1480, 1775-2000, and >2400 nm. Averaged spectra (n=32).

A modified continuum removal technique was developed for the wetland vegetation spectra following methods outlined in Schmidt and Skidmore [10]. The modification was necessary because some reflectance peaks can disregard relatively smaller, yet potentially important reflectance peaks. For example, the reflectance peak associated with biomass for green vegetation can have reflectance in the near-infrared an order of magnitude higher than that found in the visible portion of the spectrum. Therefore, a piecewise or 'modified continuum removal' was applied to the plant spectra. Specifically, a modified convex hull was forced, or fit, to include seven primary spectral reflectance maxima distributed among separated spectral regions. Once the modified continuum removal was fit, the continuum was removed by dividing the reflectance by the convex hull [16]. The seven primary spectral regions isolate the major reflectance features and were:

- Visible domain and chlorophyll absorption region (350-675 nm)
- Red edge (676-780 nm)
- Near-infrared plateau (781-975 nm)
- Near-infrared down slope (976-1190 nm)
- Upper near-infrared shoulder (1191-1450 nm)
- First shortwave infrared plateau (1451-2000 nm)
- Second shortwave infrared plateau (2001-2400 nm)

The second derivative analysis applied in this research was conducted to characterize potentially unique wavelength locations of absorption and reflection features within the collected spectra. In this research, a MATLAB script was created that fit a piecewise cubic spline to smooth a non-continuous/unsmoothed spectra in order to create a polynomial from which true second derivative values could be calculated at each band location. The five highest magnitude positive and negative values were selected to identify wavelengths possessing distinct diagnostic spectral change. This percentage was chosen because inspection of the data shows that derivatives and their paired wavelengths resulting from inflection points caused by system noise and not botanical sources were more frequent in the "middle" 80% of the data. The high magnitude values represent points of inflection that are located at the center of a reflectance (negative values capture convex features) and/or absorption feature (positive values capture concave features).

The third identification approach utilizing vegetation reflectance variation was developed by Cochrane [23] for distinguishing tropical tree species. This approach uses a *shape-filter* representing the range of reflectance variation present in a species over the spectrum. The maximum and minimum spectral reflectance for a given species for each wavelength creates its *shape-space.* If any of the comparison spectra (e.g., invasive species of interest vs. all others) fall outside the shape-space, then separation is possible. If spectra overlap for a given wavelength, then separation at that wavelength is not possible. This shape filter [23] was applied to evaluate separability of wetland invasive species in the study area. The maximum (Max) and minimum (Min) spectral reflectance at each band center/wavelength creates the shape-space for the species (Eq. 4).

$$Shape - space = Max\rho - Min\rho \text{ at each } \lambda \tag{4}$$

3. Results and Discussion

3.1. Spectral Separation

The wetland plant spectra displayed a range of JM distance values. Table 2 displays the JM distance values for all the plant spectra in a matrix against the invasive species [23].

According to the JM values, *Heteranthera dubia* (water star-grass) and *Lythrum salicaria* are relatively easy to separate with the highest JM value of 1.2645. *Heteranthera dubia* has the highest separability from most of the invasives compared to the other species collected (Table 2). The invasive with the lowest separation value from *Heteranthera dubia* is *Elodea canadensis* (another submergent species, Canadian waterweed) at 0.8599, still a moderate separation value.

Heteranthera dubia and *Elodea canadensis* are both perennial forbs with somewhat similar morphology and physiology, growing at the water-surface with a mat-like foliage texture. Both species had substantial amounts of water present in their spectra under field-canopy conditions resulting in distinguishable signatures compared to the other emergent- and more upland- aquatic invasive species in the study area. The VIS, NIR, and SWIR reflectance for *Heteranthera dubia* and *Elodea canadensis* spectra never surpassed ten percent reflectance factor because of the high amounts of water absorbing energy in the FOV.

Table 2. JM distance matrix for invasive species. *Vallisneria americana*-18 refers to a collection of spectra made with *Vallisneria americana* located approximately 18 inches below the water surface (averaged spectra n=32). (Low, medium, high detail evenly divided categories)

Wetland species	*Elodea canadensis*	*Lythrum salicaria*	*Nymphaea odorata*	*Phragmites australis*
Eleocharis rostellata	0.9258	*0.4213*	*0.1124*	*0.3245*
Elodea canadensis	-	1.066	0.9693	1.0357
Filmacutee	0.9468	0.3406	0.0969	0.2435
Heteranthera dubia	0.8599	1.2645	1.228	1.2528
Iris versicolor	0.6335	0.7902	0.5909	0.7293
Leersia oryzoides	1.1229	0.2365	0.5282	0.3381
Lemna minor	1.1257	0.2525	0.5408	0.353
Lythrum salicaria	1.066	-	0.3251	0.1093
Mowed field grass	1.0351	0.1132	0.2187	0.0051
Myriophyllum verticillatum	0.3723	0.9407	0.7983	0.8968
Nymphaea odorata	0.9693	0.3251	-	0.2225
Nuphar lutea	0.6966	0.7317	0.5134	0.6647
Phragmites australis	1.0429	0.1093	0.2225	-
Ponterderia cordata	1.0093	0.1818	0.1585	0.078
Polygonum pensylvanicum	1.0429	0.0763	0.2712	0.0643
Potamogeton spirillus	0.9264	0.418	0.1094	0.321
Sagittaria latifolia	1.0421	0.079	0.2551	0.0388
Salix eriocephala	0.9616	0.3141	0.0593	0.2127
Scirpus validus	0.7527	0.6696	0.435	0.599
Sparganium androcladum	0.6413	1.215	1.1649	1.199
Typha latifolia	0.7354	0.6946	0.4629	0.6234
Vallisneria americana	0.079	1.1444	0.9486	1.0194
Vallisneria americana-18	0.8	1.0515	1.2098	1.2371

Table 2. Continued

Wetland species	*Polygonum pensylvanicum*	*Pontederia cordata*	*Sagittaria latifolia*	*Typha latifolia*
Eleocharis rostellata	0.3663	0.2579	0.3535	0.7354
Elodea canadensis	1.0429	1.0093	1.0421	1.062
Filmacutee	0.2753	0.168	0.2672	0.3305
Heteranthera dubia	1.2552	1.2424	1.255	1.248
Iris versicolor	0.7487	0.68	0.7445	0.675
Leersia oryzoides	0.2949	0.4004	0.307	0.302
Lemna minor	0.309	0.4139	0.3218	0.4126
Lythrum salicaria	0.0763	0.1818	0.079	0.6946
Mowed field grass	0.0693	0.0761	0.0438	0.1546
Myriophyllum verticillatum	1.2182	1.202	1.218	0.8988
Nymphaea odorata	0.2712	0.1585	0.2551	0.4629
Nuphar lutea	0.6865	0.6108	0.6816	0.6578
Phragmites australis	0.0643	0.078	0.0388	0.6234
Ponterderia cordata	0.1179	-	0.1044	0.1224
Polygonum pensylvanicum	-	0.1179	0.0275	0.6474
Potamogeton spirillus	0.3625	0.254	0.3499	0.318
Sagittaria latifolia	0.0275	0.1044	-	0.0988
Salix eriocephala	0.2519	0.1389	0.2404	0.6418
Scirpus validus	0.6207	0.5378	0.6149	0.5982
Sparganium androcladum	1.2023	1.1848	1.2021	1.188
Typha latifolia	0.6474	0.5671	0.6418	-
Vallisneria americana	1.0276	0.9919	1.0264	1.0023
Vallisneria americana-18	1.2398	1.2256	1.2396	1.2217

Separability level

low medium high

Not surprisingly, *Nymphaea odorata* (water lily) had the lowest separation distance from all other species, including the invasives. This species was one of three floating species (*Lemna minor* and *Nuphar lutea*) found in this study, and typically displayed low to moderate densities. Thus, water played a large role in determining the "green" and NIR reflectance at this study site. *Sagittaria latifolia* (arrowhead) was the one invasive that displayed the lowest separability value against the other invasive species. *Sagittaria latifolia* had a relatively high separability measure from submergent invasive species such as *Elodea canadensis* at 1.0421, but very low separation scores from the other emergent/terrestrial invasives with an average of 0.1006 indicating difficulty in differentiating this species. *Sagittaria latifolia* had very small separation distances from *Polygonum pensylvanicum* (Pennsylvania smartweed) and these two are likely to cross-classify. These two plant species have relatively similar plant architectures and inhabit very similar niches in this ecosystem. The similarity of these two species with respect to the JM index lends support to the claim that plant canopy structure plays a large role in species separability. In contrast, *Sagittaria latifolia* also had very low separation values from *Phragmites australis* (common reed), an aggressive, very densely growing erect stalk with coarse texture that extends upwards of 2m, and *Nymphaea odorata*, a floating-leaf forb with large, thick circular leaves (25cm) with a significant spectra contribution from water, which suggests that plant canopy structure does not play a singularly strong role in differentiating spectra.

3.2. Continuum Removal

In theory a normalization process based on continuum removal can remove albedo, and/or background signal, from a spectral signature. In this study the modified continuum removal generally decreased separation distance as measured by the JM metric (Table 3). The differentiation of *Lythrum salicaria* from *Sagittaria latifolia*, *Potamogeton spirillus* (spiral pondweed), and *Polygonum pensylvanicum* increased slightly, although these species had very high JM separation values before the continuum removal was applied. The separability of *Potamogeton spirillus* from five of the invasives also increased slightly. *Potamogeton spirillus* can be submerged or float on the water surface with long (20cm), simple leaves. Therefore, an increase in species separability from *Lythrum salicaria*, *Nymphaea odorata*, *Phragmites australis*, *Pontederia cordata* (pickerelweed), and *Sagittaria latifolia* was contrary to expected results based on plant architecture, while emphasizing the role of leaf water content influencing separation. The decrease in separability for *Elodea canadensis* versus the non-submergent species further suggests that background signal and canopy architecture were indeed disregarded via continuum removal.

The continuum removal results suggest that applying this processing technique for all wetland plant species is not useful. While continuum removal might be effective in identifying absorption feature characteristics or particular wavelengths associated with biophysical attributes, applying the technique to aid in separating plant species (or classifying image data) can be disadvantageous. Clearly, for the wetland ecosystem studied here, continuum removal decreased abilities to separate the invasives species. These results further suggest that background and canopy architecture contributes to improving the separation of wetland plant species. The results here further advocate emphasizing life form when attempting to map wetland invasive species.

The background signal, or local environment, is what often creates conditions that support hydrophytic plants. The background signal includes variations in soil moisture or water content along with understory debris and the plant residue from previous growing seasons. These background factors provide useful biophysical information that is well-known to influence spectral reflectance. Therefore, when continuum removal techniques are applied, the loss of these background signals is detrimental to spectral separation of species in many cases. In other ecosystems or applications, such as geological and mineral identification, background signal may not be useful; in wetland ecosystems they are critical.

3.3. Spectra Features from Derivatives

The second-derivative approach identified wavelengths in this investigation similar to those identified in Becker et al. [8] using a dataset from a different sampling location. Figure 2 summarizes results of the derivative approach. With little substantial inflection (little change in slope resulting in a derivative near zero) in the NIR down slope or SWIR regions, no unique wavelengths were selected from these domains with the 2nd derivative approach; nearly all wavelengths had relatively equivalent strength in these wavelengths domains. The second SWIR plateau also tended to be noisy which derivative analysis is well-known to be

Table 3. Change in JM distance values with modified continuum removal applied. Negative values indicate increase in separation values. *Vallisneria americana*-18 refers to a collection of spectra made with *Vallisneria americana* located approximately 18 inches below the water surface (averaged spectra n=32).

Wetland species	*Elodea canadensis*	*Lythrum salicaria*	*Nymphaea odorata*	*Phragmites australis*
Eleocharis rostellata	0.5615	0.3275	0.0344	0.1982
Elodea canadensis		0.7279	0.6421	0.7038
filmacutee	0.5976	0.2086	-0.0111	0.0843
Heteranthera dubia	0.8599	1.2645	1.228	1.2528
Iris versicolor	0.3549	0.6905	0.5135	0.6105
Leersia oryzoides	0.8061	0.2127	0.5114	0.2968
Lemna minor	0.8418	0.2326	0.5038	0.3382
Lythrum saicaria	0.7279		0.3017	0.0765
Mowed field grass	0.739	0.0626	0.1648	-0.0329
Myriophyllum verticillatum	0.0678	0.8029	0.6837	0.7336
Nymphaea odorata	0.6421	0.3017		0.1707
Nuphar lutea	0.4064	0.6282	0.433	0.5388
Phragmites australis	0.7134	0.0401	0.1707	
Pontederia cordata	0.7275	0.1201	0.1103	0.0087
Polygonum pensylvanicum	0.797	-0.0207	0.1822	-0.0273
Potamogeton spirillus	0.5914	-0.0948	-0.4067	-0.1708
Sagittaria latifolia	0.7686	-0.0081	0.164	-0.0274
Salix eriocephala	0.6238	0.2438	-0.0147	0.1607
Scirpus validus	0.4855	0.602	0.3898	0.5023
Sparganium androcladum	0.5946	0.9468	1.1649	1.199
Typha latifolia	0.3958	0.6254	0.4134	0.5229
Vallisneria americana	-0.0509	0.8959	0.7165	0.7411
Vallisneria americana-18	0.6076	0.6523	0.8297	0.8302

Table 3. Continued.

Wetland species	*Polygonum pensylvanicum*	*Pontederia cordata*	*Sagittaria latifolia*	*Typha latifolia*
Eleocharis rostellata	0.2157	0.1488	0.1879	0.4562
Elodea canadensis	0.797	0.7275	0.7686	0.3958
filmacutee	0.1221	0.0495	0.0859	0.2067
Heteranthera dubia	1.2552		1.255	1.2567
Iris versicolor	0.6689	0.6248	0.6279	0.5645
Leersia oryzoides	0.2194	0.3624	0.2326	0.2826
Lemna minor	0.2221	0.355	0.2523	0.3565
Lythrum saicaria	-0.0207	0.1201	-0.008	0.6254
Mowed field grass	0.0157	0.0362	0.0063	0.1058
Myriophyllum verticillatum	1.0843	1.0943	1.2034	0.8955
Nymphaea odorata	0.1822	0.1103	0.1641	0.4134
Nuphar lutea	0.5914	0.5437	0.5517	0.6245
Phragmites australis	-0.0273	0.0087	-0.0273	0.5229
Pontederia cordata	0.1179		0.0401	0.4901
Polygonum pensylvanicum		0.0748	-0.0202	0.5285
Potamogeton spirillus	0.3194	-0.232	-0.0926	-0.0264
Sagittaria latifolia	0.0078	0.0401		-0.0065
Salix eriocephala	0.2054	0.088	0.2232	0.2036
Scirpus validus	0.5166	0.4748	0.4916	0.5062
Sparganium androcladum	0.7865	1.1033	1.2021	1.1462
Typha latifolia	0.5285	0.4901	0.5072	
Vallisneria americana	0.8655	0.8042	0.8221	0.8634
Vallisneria americana-18	0.9187	0.8852	0.8752	0.6947

highly susceptible [20]. Therefore we focus the discussion on the visible to NIR portion of the spectrum up to where the first sensor transfer occurs in the ASD instrument; another problematic region for derivative techniques.

Continuum removed spectra had relatively no influence on wavelength selection via 2nd derivatives compared to unprocessed spectra. Remember that continuum removal techniques isolate features and does not change the wavelength location in which features exist. Figure 2A displays normalized (into percent) frequency of occurrence for invasives, noninvasives, and all species with their continuum removed. No unique grouping-specific wavelengths or continuum removed wavelengths were selected. The three species that receive much attention from managers for possessing overly aggressive behavior in this study region- *Phragmites australis*, *Typha latifolia*, and *Lythrum salicaria*- had similar wavelengths selected.

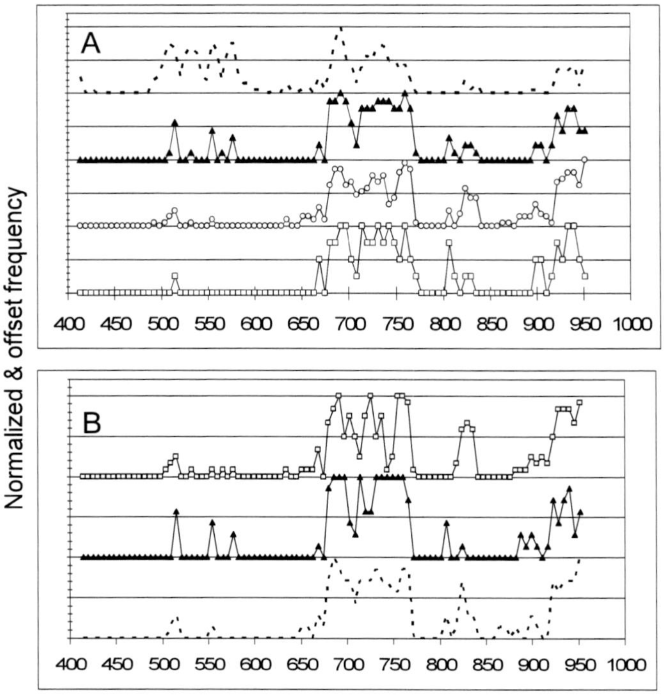

Figure 2. Frequency of large-magnitude 2nd derivative occurrences normalized into percent and offset with major axis representing units of five. Figure 2A details continuum removed 2nd derivative frequency for all species (---), grouped 2nd derivative frequency for all invasives (-▲-), grouped 2nd derivatives frequency for all noninvasives (-○-), and grouped "Big 3" (*Phragmites australis*, *Typha latifolia*, *Lythrum salicaria*) frequency (-□-). Figure 2B displays plant canopy architecture groupings for floating (-□-), shrub (-▲-), and erect orientation (---). Note the consistency of the wavelength location of the occurrences across all groupings.

Figure 2B illustrates derivative results by plant canopy architecture groups. The groupings by morphological orientation (erect, shrub, floating) had similar wavelength locations selected. These results indicate that background signal does not largely determine the wavelength location where inflection points of utility occur. The relative wavelength locations are generally consistent across canopy types and invasive status according to the 2nd derivative approach. Clearly general domain windows were identified with the NIR region representing many large-magnitude occurrences. These findings are consistence with wavelengths selected on other wetland species and other groupings [8]. The 2nd derivative analysis shows that the wavelength location of absorption and reflectance features are consistent across species and that they are largely not category (e.g., invasive or architecture) dependant.

3.4. Shape Filter

While the absorption/reflectance features from average species spectra provide useful information, in reality the reflectance for individual wetland plant spectra display considerable variation. A limitation of this study is the lack of spectra collected under varied light conditions, atmospheric conditions, and time of year. However, this investigation is the first step or pilot study to investigate if the potential exists for the techniques presented to enhance species separation in Great Lakes coastal wetlands. Figure 3 illustrates the reflectance variability for *Scirpus validus* (softstem bulrush), *Phragmites australis*, *Lythrum salicaria*, and *Typha latifolia* (broad leaved cattail). Recall that the shape filter method [23] is intended to identify species of interest, such as invasives, using the uniqueness of the absorption features and reflectance variability. In essence, the more unique an absorption feature of a given species is, the easier that species can be distinguished.

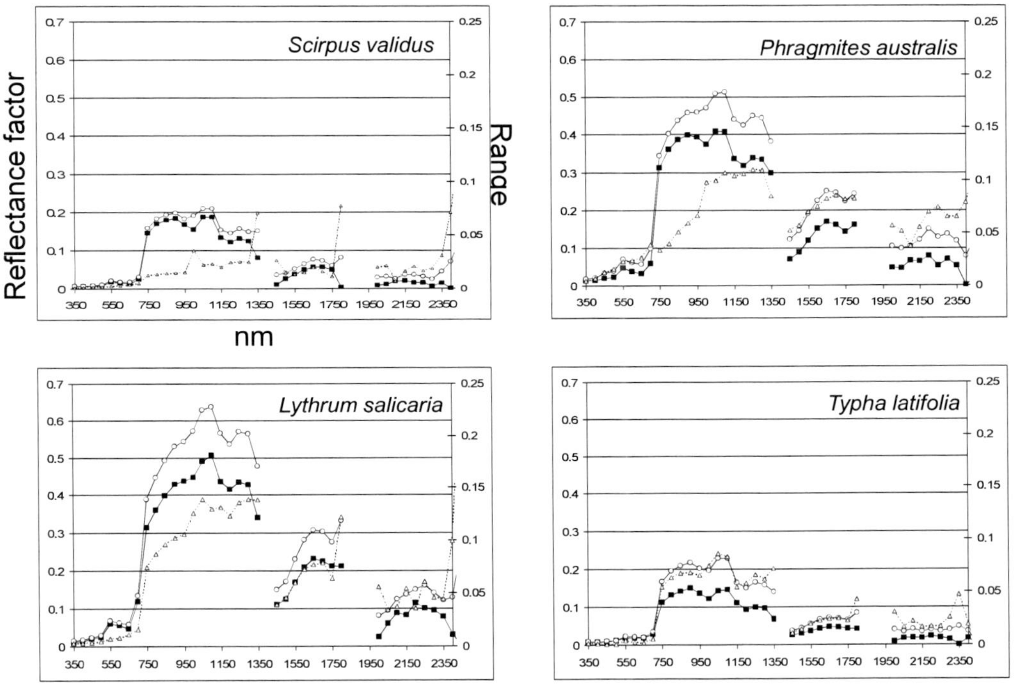

Figure 3. Reflectance factor (n=32) variability, or *shape-space*, (maximum ○, minimum ■, range -Δ-) illustrated for *Scirpus validus*, *Phragmites australis*, *Lythrum salicaria*, and *Typha australis* at 50nm intervals. Pre-processing removed wavelength regions severely affected by atmospheric absorption in the spectral ranges of 1350-1480, 1775-2000, and >2400nm.

The shape space varies by wavelength domain and by species. For example, *Scirpus validus* had relatively less variation compared to *Phragmites australis*, *Lythrum salicaria*, and *Typha latifolia*. The variation that does exist in *Scirpus validus* occurred primarily in the upper near-infrared shoulder, first SWIR plateau, and second SWIR plateau. This is likely due to the erect, small diameter structured growth and the fact that *Scirpus validus* tends to occur as a transitional plant between standing water and higher substrate on a microtopographic scale. Thus only minute differences were detected in leaf water content and plant biomass volume compared to variation in soil moisture and understory debris, again emphasizing the utility in background signal for identification purposes. *Phragmites australis* and *Typha latifolia* have larger reflectance variability in the NIR down slope (976-1190nm) and the upper near-infrared shoulder (1191-1450nm); however, both these regions have high separation abilities when the shape filter was applied (Figure 4.)

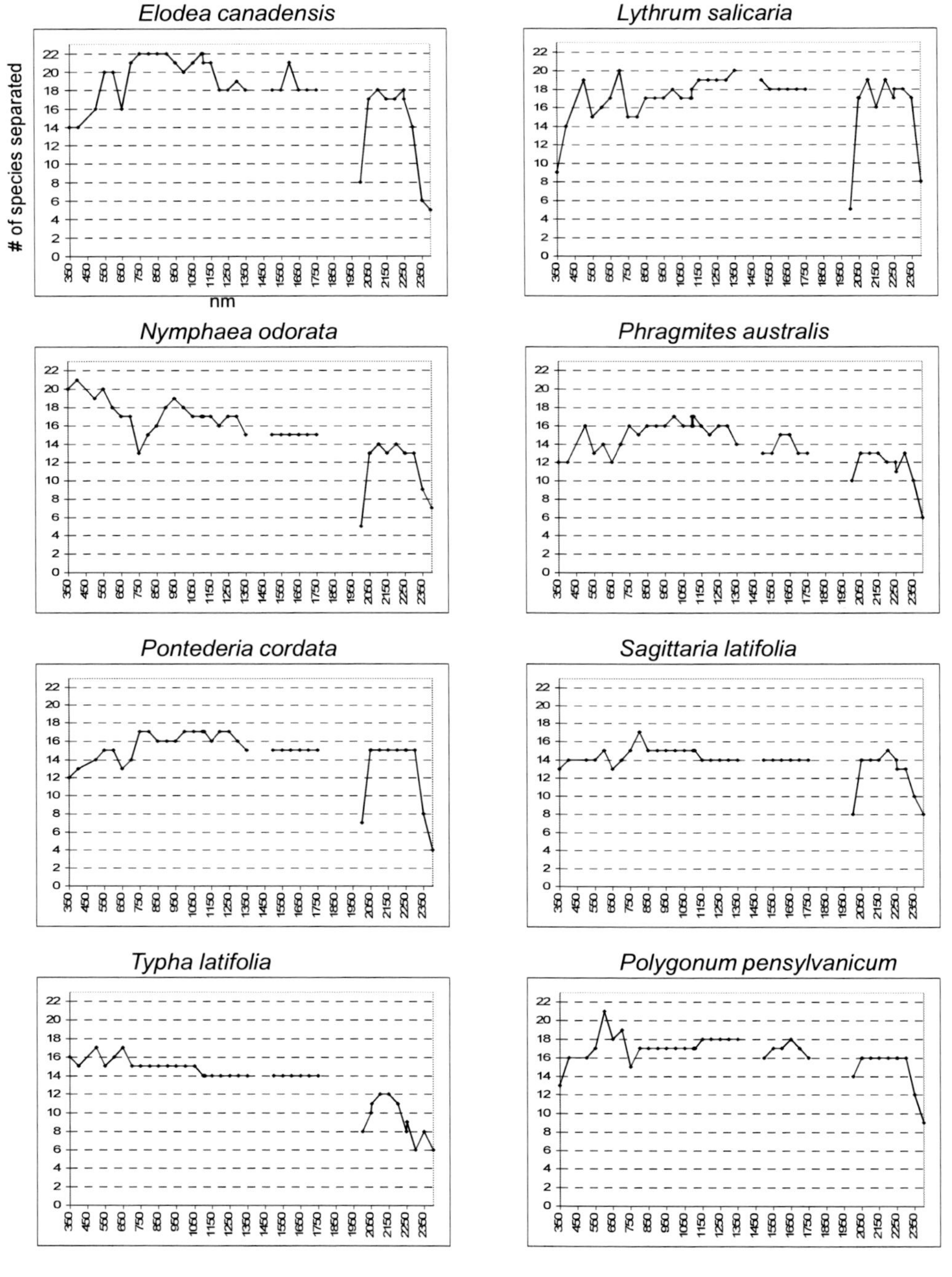

Figure 4. Number of plants species separated (displayed at 50nm intervals) by shape filtering for the wetland invasive plant species. Pre-processing removed wavelength regions severely affected by atmospheric absorption in the spectral ranges of 1350-1480, 1775-2000, and >2400nm.

The wavelengths identified as most useful for separating invasive species by the shape filter vary by species (Figure 4). This is critical as classification and processing techniques and/or choice of wavelengths might require evaluation based on the species of interest. Further, the biophysical properties influencing reflectance become valuable as background and plant canopy architecture vary by species thus potentially improving identification. For example, using the shape filter approach *Polygonum pensylvanicum* has the most separation around 605nm. Compared to the other invasives this wavelength has low separation value. *Nymphaea odorata* is most separable from the other species in the visible and chlorophyll domain (350-675nm) likely due to the species canopy being large (~24cm), round leaves that float on the water surface. Using the shape filter technique, *Lythrum salicaria* had between five (minimum at 2000nm) and 20 (maximum at 700nm) species distinguished.

The results from applying the shape filtering technique confirm that information provided by increased spectral data does increase abilities to distinguish plants of interest. The wavelength domains of utility vary by species therefore data reduction and wavelength selection methods need to consider evaluating species of interest and their individual absorption/reflectance features. The concept of spectral libraries and classification techniques based on shape filtering is promising for distinguishing invasive species. In the wetland ecosystem where this study was conducted, even very similar spectra were able to be filtered.

4. Conclusions

The focus of this study was to explore three techniques for identifying wetland invasive plant species using field-level hyperspectral data. The results are possibly applicable to airborne hyspectral systems as well, for the spectra had bandsets similar to those available from commercial airborne vendors. Continuum removal largely decreased separation according to the JM distance measure. Based on the continuum removal results, the role of canopy structure and background signal often provides great utility for separating certain wetland species. The 2nd derivative and shape filtering methods were complementary when characterizing spectra. The 2nd derivative approach found that absorption and reflectance wavelength locations were relatively consistent between species and life forms and emphasized the NIR region. Shape filtering showed that when trying to distinguish invasive species, useful wavelength regions can be species specific and method specific. This can be critical when choosing classification techniques and wavelengths as the usefulness of wavelength regions vary by species and methodology.

Acknowledgments

Funding for this project was provided in part by the Great Lakes Fisheries Trust and in part from the Society of Wetland Scientists Student Grants program. The authors also thank the entire Muskegon River Watershed Ecological Assessment Project team.

References

[1] D. Pimentel, R. Zuniga, and D. Morrison, "Update on the environmental and economic costs associated with alien-invasive species in the United States," *Ecol. Econ.*, **52**(3), 273-288(2005)[doi:10.1016/j.ecolecon.2004.10.002].

[2] W. Mitsch, and J. Gosselink, "*Wetlands*," third edition. New York: John Wiley & Sons, Inc. (2000).

[3] T. Danielson, "Methods for Evaluating Wetland Condition: Introduction to Wetland Biological Assessment." EPA 822-R-01-007a. Office of Water, U.S. Environmental Protection Agency, Washington, DC (2001).

[4] National Research Council, *Wetlands: Characteristics and Boundaries*, National Academy Press, Washington, DC (1995).

[5] National Research Council, *Ecological Indicators for the Nation*, National Academy Press, Washington, DC (2000).

[6] S. Ozesmi, and M. Bauer, "Satellite remote sensing of wetlands, "*Wetlands Ecology and Management",* **10**(5), 381-402(2002)[doi:10.1023/A:1020908432489].

[7] F. Artigas and J. Yang, "Spectral Discrimination of Marsh Vegetation Types in the New Jersey Meadowlands, USA," *Wetlands*, **26**(1), 271-277 (2006)[doi:10.1672/0277-5212].

[8] B. Becker, D. Lusch, and J. Qi, "Identifying optimal spectral bands from in situ measurements of Great Lakes coastal wetlands using second derivative analysis," *Rem. Sens. Environ*, **97**(2), 238-248(2005) [doi:10.1016/j.rse.2005.04.020].

[9] B. Becker, D. Lusch, and J. Qi., "A classification-based assessment of the optimal spectral and spatial resolutions for Great Lakes coastal wetland imagery," *Rem. Sens. Environ.*, **108**(1), 111-120(2007) [doi:10.1016/j.rse.2006.11.005].

[10] K. Schmidt, and A. Skidmore, "Spectral discrimination of vegetation types in a coastal wetland," *Rem. Sens Environ.*, **85**(1), 92-108(2003) [doi.org/10.1016/S0034-4257(02)00196-7].

[11] P. Rosso, S. Ustin, and A. Hastings, "Mapping marshland vegetation of San Francisco Bay, California, using hyperspectral data," *Int. J. Rem. Sens*, **26**(23), 5169-5191 (2005) [doi:10.1080/01431160500218770].

[12] S. Ustin, D. Roberts, J. Gamon, G. Asner, and O. Green, "Using imaging spectroscopy to study ecosystem processes and properties." *BioScience*, **54**(6), 523-534(2004) [doi:10.1641/0006-3568(2004)054[0523:UISTSE]2.0.CO;2].

[13] L. Li, S. Ustin, M. Lay, "Application of multiple endmember spectral mixture analysis (MESMA) to AVIRIS imagery for coastal salt marsh mapping: a case study in China Camp, CA, USA," *Int. J. Rem. Sens*, **26**(23), 5193-5207 (2005) [doi:10.1080/01431160500218911].

[14] E. Underwood, S. Ustin, D. DiPietro, "Mapping nonnative plants using hyperspectral Imagery," *Rem. Sens. Environ.*, **86**(2), 150-161(2003) [doi:10.1016/S0034-4257(03)00096].

[15] F. Artigas, J. Yang, "Hyperspectral remote sensing of marsh species and plant vigour gradient in the New Jersey Meadowlands," *Int. J. Rem. Sens*, **26**(23), 5209-5220(2005) [doi:10.1080/01431160500218952].

[16] R. Clark and T. Roush, "Reflectance spectroscopy: quantitative analysis techniques for remote sensing applications," *J. Geophys. Res.*, **89**, 6329-6340(1984).

[17] Z. Huang, B. Turner, S. Dury, I. Wallis, and W. Foley, "Estimating foliage nitrogen concentration from HYMAP data using continuum removal analysis," *Rem. Sens. Environ.*, **93**(1-2), 18-29(2004)[doi:10.1016/j.rse.2004.06.008].

[18] R. Kokaly, D. Desoain, R. Clark and K. Livo, "Mapping vegetation in Yellowstone national Park using spectral feature analysis of AVIRIS data," *Rem. Sens. Environ.*, **84**(3), 437-456(2003)[doi:10.1016/S0034-4257(02)00133-5].

[19] O. Mutanga, A. Skidmore, and H. Prins, "Predicting in situ quality in the Kruger National Park, South Africa, using continuum-removed absorption features," *Rem. Sens. Environ.*, **89**(3), 393-408(2004)[doi:10.1016/j.rse.2003.11.001].

[20] F. Tsai, and W. Philpot, "Derivative analysis of hyperspectral data," *Rem. Sens. Environ.*, **66**(1), 41-51(1998)[doi:10.1016/S0034-4257(98)00032-7].

[21] P. Thenkabail, R. Smith, and E. Depauw, "Hyperspectral vegetation indices and their relationships with agricultural crop characteristics," *Rem. Sens. Environ.*, **71**(2),158-182(2000)[doi:10.1016/S0034-4257(99)00067].

[22] P. Thenkabail, E. Enclona, M. Ashton, and B. van der Meer, "Accuracy assessments of hyperspectral waveband performance for vegetation analysis applications," *Rem. Sens. Environ.*, **91**(3-4), 354-376(2004) [doi:10.1016/S0034-4257(99)00067-X].

[23] M. Cochrane. (2000). "Using vegetation reflectance variability for species level classification of hyperspectral data," *Int. J. Rem. Sens.*, **21**(10), 2075-2087(2000) [doi:10.1080/014311600050021303].

[24] F. Danson, "Developments in the remote sensing of forest canopy structure," In F. Danson and S. Plummer (Eds.), *Advances in Environment Remote Sensing,* pp.53-69(1995).

[25] M. Laba, F. Tsai, D. Ogurcak, S. Smith, and M. Richmond, "Field determination of optimal dates for the discrimination of invasive wetland plant species using derivative spectral analysis," *Photogramm. Eng. Rem. Sens.*, **71**(5), 603- 611(2005).

[26] U.S. Department of Agriculture: Natural Resource Conservation Service, "The PLANTS Database" *National Plant Data Center,* (2006) <http://plants.usda.gov>.

[27] G. Schaepman-Strub, M. Schaepmen, T. Painter, S. Dangel, and J. Martonchik, "Reflectance quantities in optical remote sensing – definitions and case studies," *Rem. Sens. Environ.*, **103**(1), 27-42(2006) [doi:10.1016/j.rse.2006.03.002].

[28] T. Niel, T. McVicar, and B. Datt, "On the relationship between training sample size and data dimensionality: Monte Carlo analysis of broadband multi-temporal classification," *Rem. Sens. Environ.*, **98**(4), 468-480(2005) [doi:10.1016/j.rse.2005.08.011].

In: Invasive Species: Detection, Impact and Control
Editors: C.P. Wilcox and R. B. Turpin
ISBN 978-1-60692-252-1

Chapter 5

IMPACTS OF ALIEN INVASIVE PLANTS ON SOIL AND ECOSYSTEM PROCESSES IN BELGIUM: LESSONS FROM A MULTISPECIES APPROACH

Nicolas Dassonville[1], Sonia Vanderhoeven[2], Sylvie Domken[1], Pierre Meerts[1] and Lydie Chapuis-Lardy[3]

[1] Laboratoire de Génétique et Ecologie végétales, Université Libre de Bruxelles, 1850 chaussée de Wavre, 1160 Bruxelles, Belgium

[2] Laboratoire d'Ecologie, Faculté Universitaire des Sciences Agronomiques de Gembloux. 27, Avenue Maréchal Juin, 5030 Gembloux, Belgium

[3] UR SeqBio, Institut de Recherche pour le Développement (IRD), BP 434, 101 ANTANANARIVO, Madagascar

ABSTRACT

We have examined impacts of alien invasive plants on soil chemical properties, primary productivity and nutrient cycling in the plant / soil system. Specifically, we tested if impacts follow a general pattern across sites and species or, alternatively, if they are entirely idiosyncratic. The study first focused on 36 sites in Belgium invaded by one of the 7 most invasive plant species in NW Europe (*Solidago gigantea, Fallopia japonica, Senecio inaequidens, Heracleum mantegazzianum, Impatiens glandulifera, Prunus serotina* and *Rosa rugosa*). We compared invaded to adjacent uninvaded plots for selected parameters. Primary productivity and nutrient uptake were always higher in invaded stands compared to uninvaded plots. Magnitude and direction of impacts on soil chemical properties strongly varied depending on site. However, impacts followed a general pattern, being predictable from soil chemical properties prior to invasion. Thus, in sites with low soil nutrient contents, invasion tended to increase available nutrient pools in the topsoil while the opposite trend was observed in soils initially rich in nutrients. This suggests that exotic plant invasion could lead to the homogenization of soil nutrient concentrations across invaded landscapes.

Later on, we selected two species (*Solidago gigantea* and *Fallopia japonica*) to study in details the mechanisms of the impacts on soil properties.

In soil invaded by *S. gigantea*, soil phosphorus availability was increased. Higher turnover rates of phosphorus in belowground organs and mobilization of soil sparingly soluble P forms through rhizosphere acidification may be involved in the observed differences in soil P status between invaded and uninvaded plots.

In grassland invaded by *Fallopia japonica*, the carbon and nitrogen cycling were deeply modified. Due to its higher lignin/N ratio compared to resident vegetation, *Fallopia* litter decomposed much more slowly and immobilized a large amount of inorganic N, reducing the availability of this element in soil. On the other hand, the internal cycling of N in *Fallopia* was found exceptionally efficient. Indeed, about 80 % of the N present in aboveground biomass in summer is translocated to the rhizomes before leaves abscission. This process makes the plant relatively independent from soil N mineralization and possibly contributes to the high productivity and invasive success of the species. In addition, *F. japonica* also impacted soil fauna communities. The density of invertebrates under the canopy of *F. japonica* was reduced and the composition of the community shifted from a typical grassland community to typical forest groups. These changes may be explained by a reduction of food diversity, a change in soil microclimate and in organic matter quality.

Introduction

The strong negative impact of alien invasive species (AIS) on indigenous biodiversity has long been acknowledged by biologists (Braithwaite et al. 1989; Pysek & Pysek, 1995 ; Dunbar & Facelli, 1999; Belnap & Philips, 2001 ; Alvarez and Cushman 2002; Maertz et al., 2005 and many others). Beyond their impact on plant and animal communities, invasive plants are also able to deeply modify the functioning of invaded ecosystems (Ehrenfeld 2003; Liao et al. 2008 and references herein). Invasive plants modify the functioning of ecosystems because they differ from indigenous vegetation by several important functional traits (Wilsey and Polley 2006; Liao et al. 2008). Most of the time, they have a higher productivity than the vegetation they invade and their leaves have a higher nutrient concentration. These two characteristics often result in an accelerated turnover of nutrients and a higher availability of nutrients in invaded soils (Musil 1993; Scott et al. 2001; Duda et al. 2003; Vanderhoeven et al. 2005; Chapuis-Lardy et al. 2006; Liao et al. 2008). While most studies on alien invasive plants report increased carbon and nutrient concentration in invaded habitats, others report the opposite pattern (Christian and Wilson 1999; Leary et al. 2006). Several authors have pointed out the high variability of response of the ecosystem to invasion (Ehrenfeld, 2003; Liao et al. 2008). However, the role of the environment ("site factor") in this variability has been somewhat neglected. Most studies are "species-oriented" and focus on only one site. However, the few authors who worked in several sites pointed out that this variability can even be observed within a single species. For instance, Scott et al. (2001) observed that *Hieracium pilosella* tends to increase soil organic matter content in some sites and do the opposite in other sites without being able to explain the differences. While it was commonly admitted that the impact of an invasive species could vary depending on the environment (Ehrenfeld, 2003), no general pattern was identified until our study on the impact of invasive plant species in Belgium (Dassonville et al., 2008).

INVASIVE PLANTS TEND TO HOMOGENIZE SOIL PROPERTIES ACROSS INVADED LANDSCAPES

This first study was designed to explore the global pattern of impact of seven of the most invasive plant species in NW Europe (*Impatiens glandulifera* (*Balsaminaceae*, annual), *Heracleum mantegazzianum* (*Apiaceae*, hemicryptophyte), *Senecio inaequidens* (*Asteraceae*, chamaephyte), *Solidago gigantea* (*Asteraceae*, perennial rhizomatous geophyte), *Fallopia japonica* (*Polygonaceae*, perennial rhizomatous geophyte), *Rosa rugosa* (*Rosaceae*, woody shrub) and *Prunus serotina* (*Rosaceae*, tree)) on productivity and soil chemical properties. To take into account the "site factor", for each species, 5 invaded sites were selected in a wide variety of ecosystems across Belgium. In each site, we compared topsoil (0-10 cm) chemical properties (pH, exchangeable cations and P concentrations and organic C and N concentrations) and biomass production between invaded plots and area with the remaining uninvaded resident vegetation (For technical details, see Dassonville et al., 2008 and Vanderhoeven et al., 2005). Therefore, the impact of an invasive species at one site on one parameter was defined as the difference between the value of this parameter in invaded and uninvaded stands.

Contrary to other studies, we did not find a systematic positive impact of the invasive plants on topsoil carbon and nutrient pools. On the other hand, we found a strong correlation of the magnitude and direction of impact with initial nutrient pool in uninvaded plots. This correlation was significant for exchangeable K, Mg, Mn and P and for organic nitrogen concentration (figure 1). In oligotrophic and mesotrophic habitats, our results were in agreement with earlier reports (Musil 1993; Scott et al. 2001; Duda et al. 2003; Vanderhoeven et al. 2005; Chapuis-Lardy et al. 2006; Liao et al. 2008). The nutrient concentration of topsoil was increased consecutively to the invasion (higher nutrient concentration in the invaded soil compared to the soil under the surrounding resident vegetation). However, the magnitude of impact decreased with initial nutrient pool in topsoil and even became negative (lower concentration in the topsoil of invaded plots compared to uninvaded ones) in the most eutrophic sites. Opposite impacts of alien invasive species in nutrient-poor vs. nutrient-rich sites thus result in convergence of invaded plots towards similar values of topsoil chemical properties. Thus, element concentrations in topsoil varied within narrower limits in invaded (I) plots compared to uninvaded (U) ones, most strikingly so for Mn (44-fold variation among U plots vs. 15-fold in I plots), K (45-fold vs. 22-fold), P (40-fold vs. 22-fold), Mg (35-fold vs. 23-fold), N (32-fold vs. 21-fold), Zn (248-fold vs. 174-fold). This is not such a surprising result considering that invasion results in convergence of plant community composition by the replacement of a wide variety of ecosystems by a few dominant species with common functional traits. This study was however the first evidence of homogenization of soil properties across invaded landscapes.

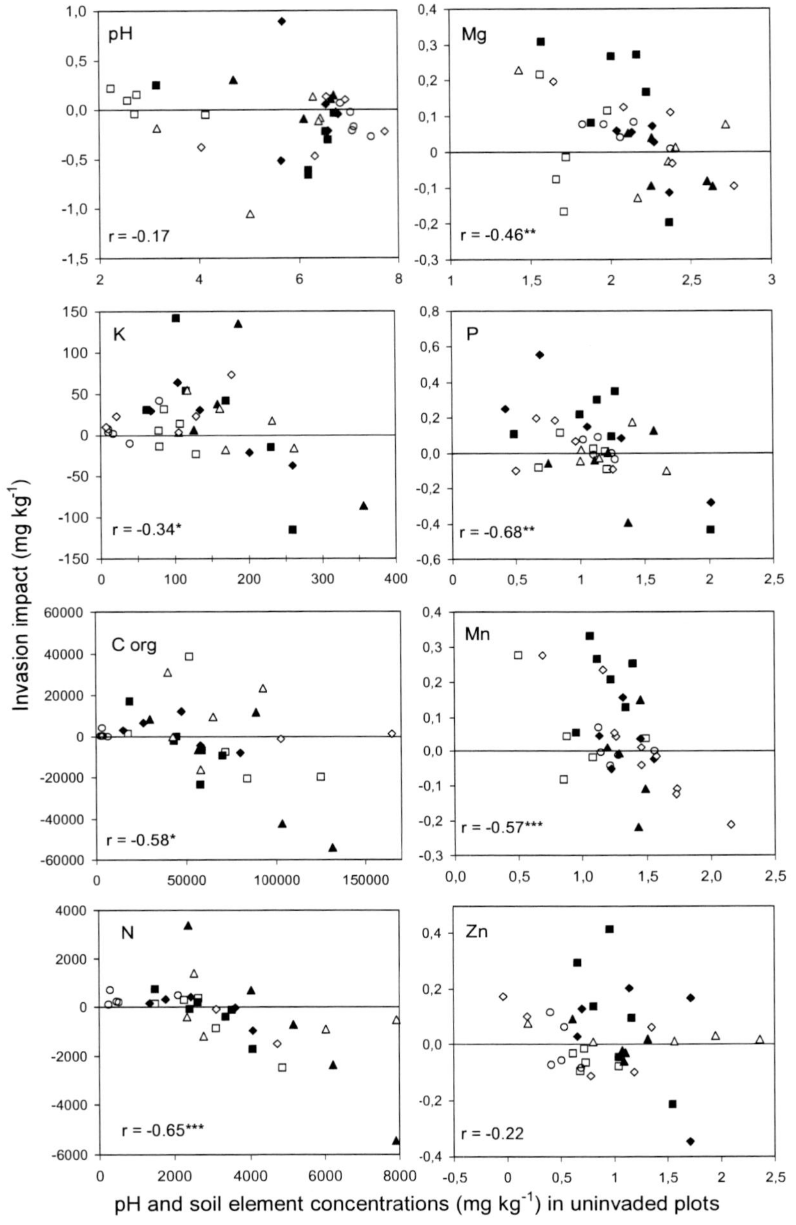

Figure 1. Impact of invasive species on soil chemical properties (pH, exchangeable cations and P concentrations and organic C and N concentrations) as a function of soil chemical properties in uninvaded plots (U). Impact is calculated as the difference between the value of the parameter in invaded plots (I) minus the value of the parameter in uninvaded plots (U). Positive values of I-U indicate increased element concentration in the invaded soil. Pearson correlation coefficient (r) between U and I-U. Significance levels (*: p<0.05, **: p<0.01, ***: p<0.001). Log-transformed data for Mg, P, Mn, Zn. ♦ *S. gigantea*, ■ *F. japonica*, ▲ *H. mantegazzianum* , Δ *I. glandulifera*, ○ *R. rugosa*, ◊ *S. inaequidens*, □ *P. serotina*. Figure from Dassonville N., Vanderhoeven S., Vanparys V., Hayez M., Gruber W., Meerts P., 2008. Impacts of alien invasive plants on soil nutrients are correlated with initial site conditions in NW Europe. Oecologia, DOI : 10.1007/s00442-008-1054-6. Copyright Springer-Verlag Heidelberg. Reproduced with permission.

MECHANISMS OF INVASIVE PLANT IMPACT ON SOIL PROPERTIES

The mechanisms implied in these soil properties modifications are probably diverse and are not completely understood. However, our data allow us to formulate some hypotheses. The biomass production of all studied plants was always higher than that of the resident vegetation (Dassonville et al., 2008). Higher biomass was achieved either by larger shoots and faster growth rate (*F. japonica, I. glandulifera, S. gigantea*) or by increased ground cover compared to the resident vegetation (*S. inaequidens*). Despite lower or equal nutrient concentration in aboveground biomass of the invasive plants compared to uninvaded vegetation, the nutrient standing stocks were always higher due to the higher biomass (Dassonville et al., 2008). These higher aboveground nutrient stocks can be achieved by the invasive plants probably through access to nutrient sources inaccessible for the resident vegetation. Nutrient uplift (*sensu* Jobbagy & Jackson, 2004) by a deeper root system is certainly the explanation for *Fallopia japonica* which can root at more than 2 meter deep (Dassonville et al., 2007). This is also the case for *H. mantegazzianum* and probably also for *R. rugosa* and *P. serotina* surrounded by herbaceous vegetation. Active mobilisation of nutrient forms that were inaccessible for resident vegetation is also possible as shown by Chapuis-Lardy et al (2006) for P. In absence of a strong translocation, this leads to a higher nutrient return to the soil through litter fall. Thus the invasive species steadily enhance specific nutrient fluxes in invaded ecosystems, including net uptake rates from soil and annual returns in dead organic matter. This higher nutrient return could explain the increased nutrient concentrations in the topsoil of the originally most oligotrophic sites. However, it must be strong enough to compensate the increased nutrient leaching in invaded plots. Invaded plots are more sensible to nutrient leaching because of the absence of soil cover during nearly 6 months in winter and early spring contrary to the resident vegetation, mostly composed of perennial grass species (Hooper and Vitousek 1998; Ridley et al. 2001). This phenomenon may explain why the magnitude of impact decreases with increased initial nutrient concentration in the soil. In addition, in the most fertile sites, nutrients may be less strongly immobilised in the topsoil compared to nutrient-limited sites (Tamm et al. 1995) and result in "nutrient leakage". Organic matter turnover rates and nutrient leaching losses are often increased in eutrophic sites (Wedin and Tilman 1996; Debusk and Reddy 1998).

We essentially discussed the "site factor" which in our opinion explains a high proportion of the variability of impact (Dassonville et al., 2008). However, all studied species differ in their strategies and even for two plants (*S. gigantea* and *F.* japonica) with the same growth form (perennial rhizomatous geophytes), we found some divergence in their impacts. For these reasons, the impacts of these two species were studied in more details.

SOIL PHOSPHORUS IN ECOSYSTEMS INVADED BY GIANT GOLDENROD (*SOLIDAGO GIGANTEA*)

As reported in Dassonville et al. (2008), the main impact of *Solidago gigantea* was on soil P. In most studied sites, P availability increased consecutively to invasion. Nutrient uplift is not the mechanism responsible for this increased as the rooting depth is comparable in both invasive and resident vegetation (Herr et al., 2007). In order to understand the mechanisms affecting P availability, we examined seasonal variation of labile inorganic and organic P fractions, microbial P, and phosphomonoesterase activity in soil as well as P in standing biomasses (above- and belowground) (Herr et al, 2007). Higher concentrations of several labile fractions of P were observed in invaded soils compared to uninvaded soils (figures 2 and 3). Soil pH was also lower in invaded stands than under the resident vegetation. Soil pH is one of the most important parameters determining adsorption/desorption equilibrium of phosphates in soils (Hinsinger, 2001). The observed half-a-unit decrease of pH might favour solubility of mineral P compounds as suggested by *Chapuis-Lardy et* al. (2006). Decreased pH was also invoked as a possible mechanism for the increased P availability under the exotic invasive *Lepidium latifolium (Blank* and *Young,* 2002). The P stock in aboveground biomass of *Solidago gigantea* is higher than in the resident vegetation so that the quantity of P returning to the soil through litterfall is slightly higher in invaded plots. However, these amounts are not sufficient to explain a rapid increase in P availability in invaded plots compared to uninvaded ones. Fast nutrient leaching and mineralization were already reported for roots, suggesting that roots play a major role in the cycling of nutrients (Scheffer and Aerts, 2000). The increased P availability under *Solidago gigantea* probably comes mainly from the decomposition of belowground dead biomass. A massive P uptake takes place in autumn alimenting the production of fine roots. This P accumulation exceeds the P needs for aboveground biomass production in the subsequent year. Then, due to massive fine roots death in winter and spring, most of the P stored in belowground organs in autumn is lost in belowground litter. Phosphorus is easily released from dying roots (Eason and Newman, 1990) and quick release of P from decomposing roots was invoked as a possible mechanism of P enrichment in the soil (Campbell et al., 1993). Intense mobilization of P by *Solidago* roots in autumn and restitution of easily decomposable root debris in next spring may contribute to the higher availability of soil P observed in invaded plots in our study.

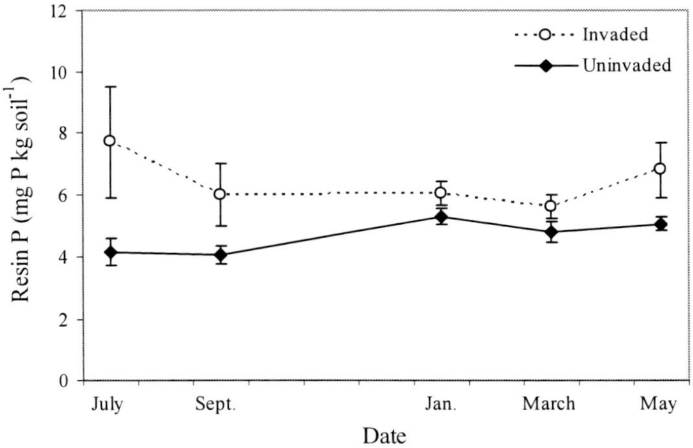

Figure 2. Seasonal variation of soil resin-Pi (0-10 cm) in plots invaded by early Goldenrod and adjacent, uninvaded plots. Means (n=6) and standard deviations. Figure from Herr C., Chapuis-Lardy L., Dassonville N., Vanderhoeven S. & Meerts P., 2007. Seasonal effect of the exotic invasive plant *Solidago gigantea* on soil phosphorus pools and fluxes. Journal of Plant Nutrition and Soil Science 170: 729-738. Copyright Wiley-VCH Verlag GmbH & Co. KGaA. Reproduced with permission.

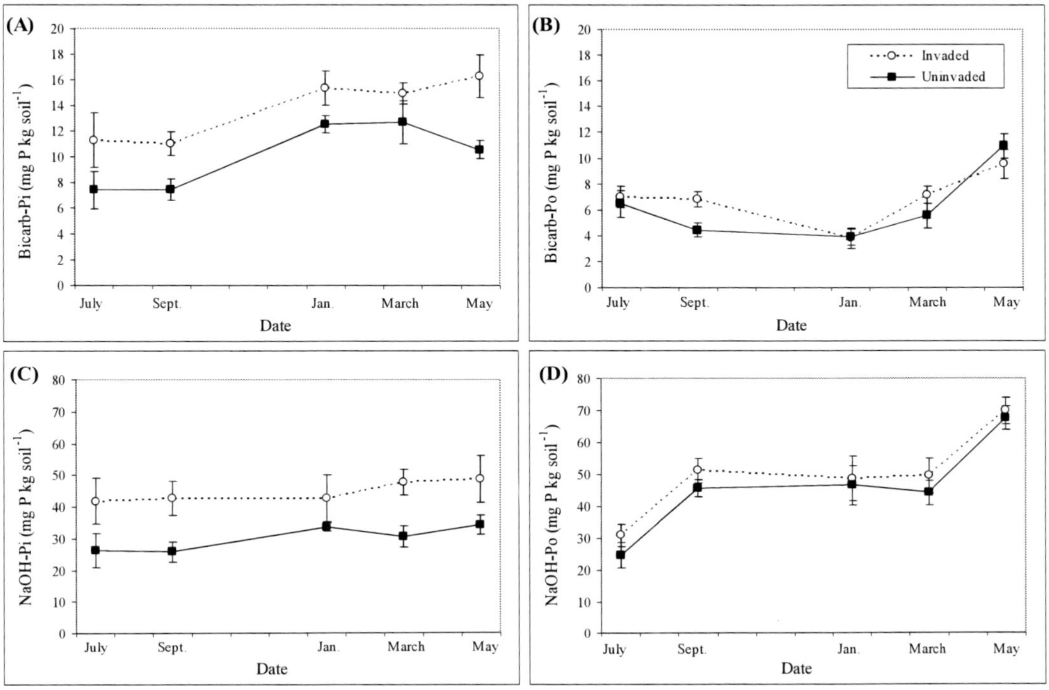

Figure 3. Seasonal variation of different soil P fractions (0-10 cm) in plots invaded by early Goldenrod and adjacent, uninvaded plots. (A) bicarb-Pi, (B) bicarb-Po, (C) NaOH-Pi, (D) NaOH-Po. Means (n=6) and standard deviations. Figure from Herr C., Chapuis-Lardy L., Dassonville N., Vanderhoeven S. & Meerts P., 2007. Seasonal effect of the exotic invasive plant *Solidago gigantea* on soil phosphorus pools and fluxes. Journal of Plant Nutrition and Soil Science 170: 729-738. Copyright Wiley-VCH Verlag GmbH & Co. KGaA. Reproduced with permission.

Alteration of Litter Decomposition and Nitrogen Cycling by Japanese Knotweed

In Dassonville et al. (2008), the Japanese knotweed (*Fallopia japonica*) appeared as the species with the strongest impacts on soil nutrients (especially cations and P), probably after a nutrient uplift from deeper horizons. On the other hand, surprisingly, N is not concerned by such a nutrient uplift process as total N concentration in the Ah horizon tended to be unchanged (Dassonville et al., 2007). Nitrogen concentrated in the topsoil as soil organic matter presents a large reserve of nitrogen. Higher biomass and litter production should have an impact on N and organic matter dynamic. Therefore, we examine in more details litter decomposition and nitrogen cycling in monospecific stands of *Fallopia japonica* as well as the impacts of the invasive on soil fauna involved in decomposition processes. Native vegetation was a rough grassland dominated by *Calamagrostis epigejos* and *Eupatorium cannabinum* (details on the study site (Ghi) in Dassonville et al., 2007).

Litter decomposition is controlled by three main factors: litter quality, structure of decomposing organism communities and microclimate (Coûteaux et al., 1995). The invasion of a plant community by a novel species can potentially modify these three factors. A litterbag experiment with replicates at random was set up in the field to measure the differences of mass loss rate and N dynamics between indigenous (50-50 mixture of *Calamagrostis epigejos* and *Eupatorium cannabinum*) and *Fallopia* litter, and to assess the impact of the microclimate and the role of soil fauna community (invaded versus uninvaded environment) in the litter decomposition process.

LITTER QUALITY

The litters collected in *Fallopia* and native vegetation stands were chemically very different (table 1). Senescent *Fallopia* leaves and stems had extremely low N concentrations (0.63 and 0.30 % respectively), and accordingly high C/N ratios (72 and 151 respectively) while the indigenous litter mixture had a low C/N ratio (33) due to a high N concentration (1.38 %). The *Fallopia* litter C/N ratio is exceptionally high compared to most deciduous broadleaves species and even conifer tree species. For instance, litter of *Fagus sylvatica* and *Picea abies*, which are known to have very slow decomposing litter, have a C/N ratio of 56 and 49 respectively (Hobbie et al., 2006), which is still lower than that of *Fallopia* organs. On the other hand, the C/N ratio of the indigenous litter mixture is lower than that of *Tilia cordata* (37), which is known to have a particularly rapid decomposing litter (Hobbie et al., 2006). The C/N ratio found in the native litter mixture is comparable to that of typical temperate grassland litter. Whitehead (1995) found that C/N ratio in dead grass herbage varies between 19 in N fertilized grassland to 44 in grassland with no or moderate N fertilization. The lignin concentration is also a good predictor of litter decay rate (Funk et al., 2005). In *Fallopia* organs, it was higher compared to indigenous leaves (14.5 % in *Fallopia* leaves and 18.8 % in stems vs. 12.1 % in indigenous mixture, table 1). Compared to indigenous herbaceous species, *Fallopia* has thus a low quality litter. For instance, *Poa pratensis* has only 5.7 % lignin in its senescent leaves and *Bromus tectorum*, which is known to decrease litter decomposition rate in the Great plains of North America has 7.4 % lignin in its senescent leaves (Ogle et al., 2003).

Table 1. Chemical composition of the three litter types

Litter type	C (%)	N (%)	C/N	Lignin (%)	Lignin/N
Fallopia leaves	45.7 ± 0.2	0.63 ± 0.01	72.5 ± 1.2	14.53 ± 0.04	23.1 ± 0.4
Fallopia stems	45.4 ± 0.3	0.30 ± 0.01	151.3 ± 5.1	18.82 ± 0.05	62.7 ± 2.1
Indigenous mixture	45.5 ± 0.2	1.38 ± 0.02	33.0 ± 0.5	12.07 ± 0.03	8.7 ± 0.1

Mean ± standard deviation (N=2 analytical repetitions for C, N, lignin and cellulose and 3 for Ca, Mg, K and P)

According to its lower quality, *Fallopia* litter decomposed more slowly than indigenous litter (figure 4). After one year, the indigenous litter lost *circa* 90 % of its initial mass while *Fallopia* leaves and stems lost only 39 and 55 % of their initial mass, respectively. Moreover, contrarily to Fallopia, indigenous litter was more rapidly attacked by decomposers, the structure of the litter being hardly recognizable after only a few weeks. Some invasive species (mostly grass species) have been shown to also have a recalcitrant litter with a slow decomposition rate (Ehrenfeld et al., 2001, Ogle et al., 2002; Drenovsky & Batten, 2007). For instance, *Aegilops triuncalis* litter have a higher lignin/N ratio and decomposed much more slowly than the litter of the native grassland species (Drenovsky & Batten, 2007). However, many other studies reported (Ehrenfeld et al., 2001; Allison & Vitousek, 2004; Rothstein et al. 2004; Standish et al., 2004; Ashton et al., 2005; Funk, 2005; Hughes and Uowolo, 2006...) higher litter quality (higher N concentration, lower C/N ratio, lower lignin/N ratio) and faster decomposition rates for invasive compared to native species.

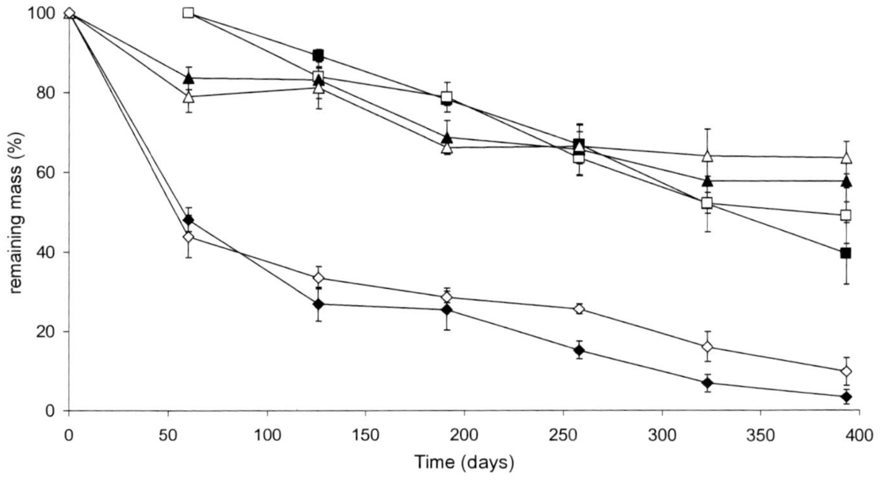

Figure 4. Decomposition of *Fallopia* leaves (triangles) and stems (squares) and indigenous litter (diamonds) during one year (between October 2006 and October 2007). All litter types were incubated in invaded (black) and uninvaded (white) environments; Decomposition is expressed as the percentage of initial mass loss Values are means ± standard deviation.

The slow decomposition rate of *Fallopia* compared to indigenous mixture may be accounted for by its much lower quality (higher C/N and lignin/N ratios). Indigenous litter could also have benefited for a "mixture effect". Indeed, in their review on litter mixture decomposition, Gartner and Cardon (2004) found that, most of the time (67 % cases) mixtures decompose faster than expected with additive decomposition rate of the components of the mixture. The synergistic effect of mixture can be explained by nutrients transfer between nutrient rich and nutrient poor components. For instance, in a mixture experiment, the litter of the N fixing *Vicia lathyroides* helped the decomposition of the less decomposable (N poorer) *Calamagrostis epigejos* (Hoorens et al., 2002). A similar mechanism could explain the high decomposition rate found in our native mixture. The relatively N rich *Eupatorium* accelerated the decomposition of *Calamagrostis*.

The C/N ratio of indigenous litter did not vary during decomposition and the N release followed the decomposition curve (figure 5) indicating that N is not limiting for the decomposers community development. On the other hand, N concentration in *Fallopia* litter increased from 0.6 % at the beginning of the experiment to 1.4 % after one year. This increased N concentration was not due to a corresponding C loss. Rather, it is most likely the result of net N fixation by the decomposers community (figure 5). Gallardo and Merino (1992) showed that litter N concentration increased during decomposition until litter C/N ratio was equal to that of the decomposers. Lignin is, in this case, a sink for N, forming the precursors of humus.

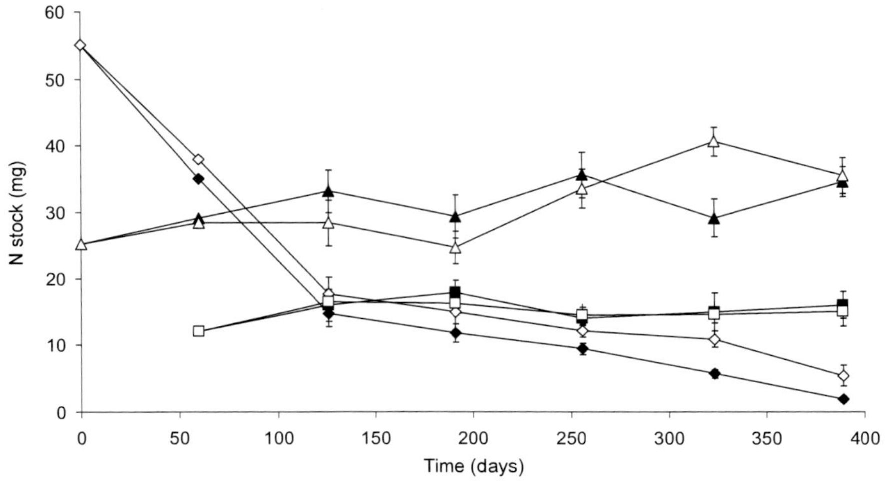

Figure 5. Evolution of the N stock (=remaining mass x N concentration) in *Fallopia* leaves (triangles) and stems (squares) and indigenous litter (diamonds during decomposition in invaded (black) and uninvaded (white) environment. Values are means ± standard deviation.

IMPACT OF *FALLOPIA* ON SOIL FAUNA

Many invasive plants have been shown to modify the structure of soil fauna and microflora communities (Wardle *et al.*, 1995; Belnap & Philips, 2001; Kourtev *et al.*, 2002; Standish *et al.*, 2004; Herrera & Dudley, 2003; Ernst and Cappuccino, 2005; Li *et al.*, 2006…). In our study, the total number of individuals under *Fallopia japonica* canopy was 50% lower ($p<0.05$) than in the uninvaded adjacent vegetation. These results are in agreement with those of Gerber et al. (2008) who studied the impact of invasive knotweeds in European riparian environments. A multivariate analysis allowed to clearly separate invaded and uninvaded plots in terms of fauna taxonomic groups (figures 6 and 7).

Moreover, faunistic assemblage was more homogenous in invaded than in uninvaded plots accordingly to the reduction of plant and litter diversity that contributed to decrease microhabitats diversity and resources heterogeneity (Dennis et al, 1998 ; Hassan, 2000 ; Oliver et al, 2000, Haddad et al, 2001). Despite a lower total density of soil fauna in invaded plots, we found an increase for some groups such as woodlice, millipedes and earthworms that play an important role in litter decomposition. The impact of *Fallopia japonica* on epigeic fauna has also been studied by Kappes et al. (2007) who found less herbivores and more carnivores (opilones) in invaded stands but an equivalent detritivorous fauna. In our study, some groups with affinity for shadow and humid environments (e.g., Isopods or diplopods) were more abundant under *Fallopia* while thermophilous organisms such as the ant *Lasius flavus* and the associated aphids were totally absent from invaded plots. Another interesting result concerned the differences in earthworms: species frequently observed in moist environment (i.e.: *Lumbricus terrestris, Dandrobaena subribicunda*) were only present in the invaded plots while grassland typical species (*Lumbricus castaneus*) (Bouché, 1972) were restricted to soil under the resident vegetation. Thus, soil fauna tend to shift from a typical herbaceous vegetation fauna to a typical forest fauna. Topp et al. (2008) who worked on the impact of *Fallopia* on soil dwelling beetles had the same conclusion. These changes in soil fauna communities should have contributed to the higher litter decomposition rate in invaded plots.

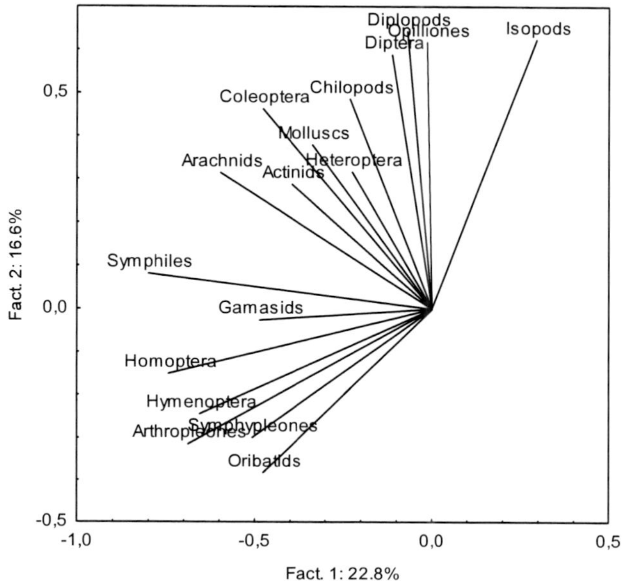

Figure 6. Soil fauna in a site invaded by *Fallopia japonica*. Factor coordinates of the variables (taxonomic groups) in the PCA based on correlations.

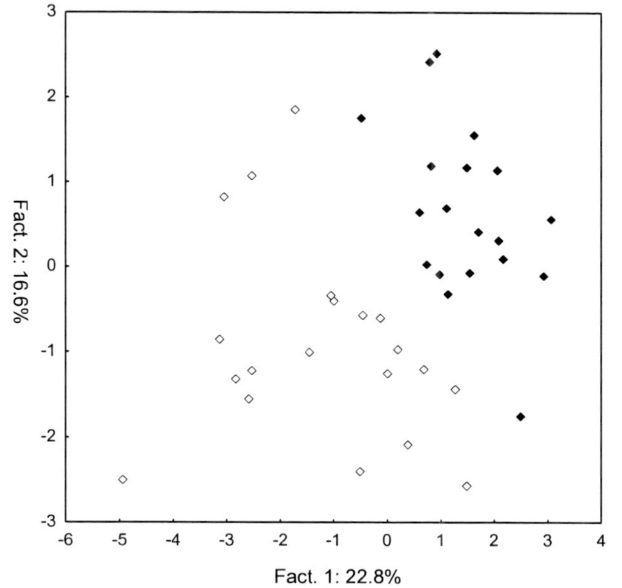

Figure 7. Soil fauna in a site invaded by *Fallopia japonica*. Factor coordinates of cases (soil plots) in the PCA. Invaded plots (black) and uninvaded plots (white).

Changes in Microclimate

By changing the vegetation structure, invasive species can influence microclimate. For instance, the higher litter decomposition rate in New Zealand podocarp forest invaded by *Tradescantia fluminensis* was attributed to the wetter microclimate in dense mats of the invasive species (Standish et al., 2004). Similarly, in Australian dunes invaded by *Chrysanthemoides monilifera* litter decomposed faster under the invasive plant than under the native vegetation probably because *Chrysanthemoides* provided wetter conditions (Lindsay & French, 2004).

In our experiment, the decomposition rate was slightly but significantly higher in invaded than in uninvaded plots (figure 4). This difference may also be linked to wetter conditions in the invaded plots as revealed by the development of faunal communities which usually lived in moist environments.

Conclusion

Our research highlighted the variability of impact of alien invasive plant species on soil and ecosystem processes. We showed that this variability has two components. The first component is what we called the "site factor" which represents the role of the environment in determining the direction and amplitude of impact. For the seven species tested, the impact of invasion on soil properties depended on the environment in a predictable way. The concentration of nutrients in soil increased following the invasion in initially oligotrophic sites while it decreased in more eutrophic sites. This leads to a strong homogenisation of soil properties across invaded landscapes contributing to the ongoing global ecosystem homogenization.

The second source of variability was the identity of the invasive species. Even for similar growth form, differences of impact were observed. The contrast was particularly obvious for *Solidago gigantea* and *Fallopia japonica*. The first had only an important impact on soil P while the second had an impact on all cations and P. The impact of *Solidago* is mediated by a modification of chemical equilibrium and by the mineralisation of dead fine roots and do not affect the total soil P pool. On the other hand, the impact of *Fallopia* implies the transport of nutrients from the deep soil to the topsoil through the deep rooting system of the plant. This process increases the total pool of nutrient in the topsoil. *Fallopia* can be considered as an ecosystem engineer. In addition to these chemical soil modifications, it profoundly modifies the structure and functioning of invaded ecosystems. The plant produces a high amount of slow decomposing litter in which soil mineral N is then sequestered. This, combined with a very conservative management of N, gives a competitive advantage to the invasive plant. The canopy of the plant also modifies the soil microclimate with repercussions on soil fauna which shift from a typical herbaceous community to a typical forest community.

The resilience of all these ecosystem impacts after an eventual eradication of the invasive plants should be measured.

References

Allison, S.D. & Vitousek, P.M. (2004). Rapid nutrient cycling in leaf litter from invasive plants in Hawai'i. *Oecologia*, 141, 612-619.

Alvarez, M.E. & Cushman, J.H. (2002). Community-level consequences of a plant invasion: effects on three habitats in coastal California. *Ecological Applications,* 12, 1434-1444.

Ashton, I.W.; Hyatt, L.A.; Howe, K.M.; Gurevitch, J. & Lerdau, M.T. (2005). Invasive species accelerate decomposition and litter nitrogen loss in a mixed deciduous forest. *Ecological Applications*, 15, 1263-1272.

Belnap, J. & Philips, S.L. (2001). Soil biota in an ungrazed grassland: response to annual grass (*Bromus tectorum*) invasion. *Ecological Applications*, 11, 1261-1275.

Blank, R.R. & Young, J.A. (2002). Influence of the invasive crucifer, *Lepidium latifolium*, on soil properties and elemental cycling. *Soil Science*, 167, 821-829.

Bouché, M.B. (1972). Lombriciens de France, écologie et systématique. I.N.R.A. (Annales de zoologie-écologie animale , special issue 72/2), Paris.

Braithwaite, R.W.; Lonsdale, W.M. & Estbergs, J.A. (1989). Alien vegetation and native biota in tropical Australia: the impact of *Mimosa pigra*. *Biological Conservation*, 48, 189-210.

Campbell, C.A.; Lafond, G.P.; Biederbeck, V.O. & Winkleman, G.E. (1993): Influence of legumes and fertilization on deep distribution of available phosphorus (Olsen-Pi) in a thin black chernozomic soil. *Canadian Journal of Soil Science*, 73, 555-565.

Chapuis-Lardy, L.; Vanderhoeven, S.; Dassonville, N.; Koutika, L.S. & Meerts, P. (2006). Effect of the exotic invasive plant *Solidago gigantea* on soil phosphorus status. *Biology and Fertility of Soils*, 42, 481-489.

Coûteaux, M.M.; Bottner, P. & Berg, B. (1995). Litter decomposition, climate and litter quality. *Trends in Ecology and Evolution*, 10, 63-66.

Christian, J.M. & Wilson, S.D. (1999). Long-term ecosystem impacts of an introduced grass in the northern Great Plains. *Ecology*, 80, 2397-2407.

Dassonville, N.; Vanderhoeven, S.; Gruber, W. & Meerts, P. (2007). Invasion by *Fallopia japonica* increases topsoil mineral nutrient concentration. *Ecoscience*, 14, 230-240.

Dassonville, N. (2008). Impact des plantes exotiques envahissantes sur le fonctionnement des écosystèmes en Belgique. phD thesis. Université Libre de Bruxelles, Belgium.

Dassonville, N.; Vanderhoeven, S.; Vanparys, V.; Hayez, M.; Gruber, W. & Meerts, P. (2008). Impacts of alien invasive plants on soil nutrients are correlated with initial site conditions in NW Europe. Oecologia, DOI : 10.1007/s00442-008-1054-6.

Debusk, W.F. & Reddy, K.R. (1998) Turnover of detrital organic carbon in a nutrient-impacted Everglades marsh. *Soil Science Society of America Journal*, 62, 1460-1468.

Dennis, P.; Young, M.R. & Gordon, I.J. (1998). Distribution and abundance of small insects and arachnids in relation to structural heterogeneity of grazed indigenous grasslands. *Ecological Entomology*, 23, 253-264.

Drenovsky, R.E. & Batten, K.M. (2007). Invasion by *Aegilops triuncalis* (barb goatgrass) slows carbon and nutrient cycling in a serpentine grassland. *Biological Invasions*, 9, 107-116.

Duda, J.J.; Freeman, D.C.; Emlen, J.M.; Belnap, J.; Kitchen, S.G.; Zak, J.C.; Sobek, E.; Tracy, M. & Montante, J. (2003). Differences in native soil ecology associated with invasion of the exotic annual chenopod, *Halogeton glomeratus*. *Biology and Fertility of Soils*, 38, 72-77.

Dunbar, K.R. & Facelli, J.M. (1999). The impacts of a novel invasive species *Orbea variegata* (African carrion flower) on the chenopod shrublands of South Australia. *Journal of Environments*, 41, 37-48.

Eason, W R.; Newman, E. (1990): Rapid loss of phosphorus from dying ryegrass roots: the chemical components involved. *Oecologia*, 82, 432-436.

Ehrenfeld, J.G.; Kourtev, P. & Huang, W. (2001). Changes in soil functions following invasions of exotic understory plants in deciduous forests. *Ecological Applications*, 11, 1287-1300.

Ehrenfeld, J.G. (2003). Effects of exotic plant invasions on soil nutrient cycling processes. *Ecosystems*, 6, 503-523.

Ernst, C.M. & Cappuccino, N. (2005). The effect of an invasive alien vine, *Vincetoxicum rossicum* (*Asclepiadaceae*), on arthropod populations in Ontario old fields. *Biological Invasions,* 7, 417-425.

Funk, J.L. (2005). *Hedychium gardnerianum* invasion into Hawaiian montane forest: interactions among litter quality, decomposition rate, and soil nitrogen availability. *Biogeochemistry*, 76, 441-451.

Gallardo, A. & Merino, J. (1992). Nitrogen immobilization in leaf litter at two Mediterranean ecosystems of SW Spain. Biogeochemistry, 15, 213-228.

Gartner, T.B. & Cardon, Z.G. (2004). Decomposition dynamics in mixed species leaf litter. *Oikos*, 104, 230-246.

Gerber, E.; Krebs, C.; Murrell, C.; Moretti, M.; Rocklin, R.; Schaffner, U. (2008) Exotic invasive knotweeds (*Fallopia spp.*) negatively affect native plant and invertebrate assemblages in European riparian habitats. *Biological Conservation*, 141, 646-654.

Haddad, N.M.; Tilman, D.; Haarstad, J.; Ritchie, M. & Knops, J.M.H. (2001) Contrasting effects of plant richness and composition on insect communities: a field experiment. *The American Naturalist*, 158, 17-35.

Hassan, R.A. (2000). *Diversity in the decomposing landscape*. In: Coleman, D.C. & Hendrix, P.F. (eds), Invertebrates as Webmasters in Ecosystems, CABI Publishing, pp 203–219.

Herrera, A.M. & Dudley, T.M. (2003). Reduction of riperian arthropod abundance and diversity as a consequence of giant reed (*Arundo donax*) invasion. *Biological Invasions*, 5, 167-177.

Herr, C.; Chapuis-Lardy, L.; Dassonville, N.; Vanderhoeven, S. & Meerts, P. (2007). Seasonal effect of the exotic invasive plant *Solidago gigantea* on soil phosphorus pools and fluxes. *Journal of Plant Nutrition and Soil Science*, 170, 729-738.

Hinsinger, P. (2001). Bioavailability of soil inorganic P in the rhizosphere as affected by root-induced chemical changes: a review. *Plant and Soil*, 237, 173-195.

Hobbie, S.E.; Reich, P.B.; Oleksyn, J.; Ogdahl, M.; Zytkowiak, R.; Hale, C. & Karolewski, P. (2006). Tree species effects on decomposition and forest floor dynamics in a common garden. *Ecology*, 87, 2288-2297.

Hooper, D.U. & Vitousek, P.M. (1998). Effects of plant composition and diversity on nutrient cycling. *Ecological Monographs*, 68, 121-149.

Hoorens, B.; Aerts, R. & Stroetenga, M. (2002). Litter quality and interactive effects in litter mixtures: more negative interactions under elevated CO_2? *Journal of Ecology*, 90, 1009–1016.

Hughes, R.F. & Uowolo, A. (2006). Impacts of *Falcataria moluccana* invasion on decomposition in Hawaiian Lowland Wet Forests: the Importance of Stand-level Controls. *Ecosystems*, 9, 977-991.

Jobbagy, E.G. & Jackson, R.N. (2004). The uplift of soil nutrients by plants: Biogeochemical consequences across scales. *Ecology*, 85, 2380-2389.

Kappes, H.; Lay, R. & Topp, W. (2007). Changes in different trophic levels of litter-dwelling macrofauna associated with Giant Knotweed invasion. *Ecosystems*, 10, 734-744.

Kourtev, P.S.; Ehrenfeld, J.G. & Häggblom, M. (2002). Exotic plant species alter the microbial community structure and function in the soil. *Ecology*, 83, 3152-3166.

Laskowski, R.; Niklińska, M. & Maryański, M. (1995). The dynamics of chemical elements in forest litter. *Ecology*, 76, 1393-1406.

Leary, J.K.; Hue, N.V.; Singleton, P.W. & Borthakur, D. (2006). The major features of an infestation by the invasive weed legume gorse (*Ulex europaeus*) on volcanic soils in Hawaii. *Biology and Fertility of Soils*, 42, 215-223.

Li, W.H.; Zhang, C.B.; Jiang, H.B.; Xin, G.R. & Yang, Z.Y. (2006). Changes in soil microbial community associated with invasion of the exotic weed, *Mikania micrantha* HBK. *Plant and Soil*, 281, 309-324.

Liao, C.; Peng, R.; Luo, Y.; Zhou, X.; Wu, X.; Fang, C.; Chen, J. & Li, B. (2008). Altered ecosystem carbon and nitrogen cycles by plant invasion: a meta-analysis. *New Phytologist* 177: 706-714.

Lindsay, E.A. & French, K. (2004). *Chrysanthemoides monilifera ssp rotundata* invasion alters decomposition rates in coastal areas of south-eastern Australia. *Forest Ecology and Management,* 198, 387-399.

Maertz, J.C.; Blossey, B. & Nuzzo, V. (2005). Green frogs show reduced foraging success in habitats invaded by Japanese knotweed. *Biodiversity and Conservation*, 14, 2901-2911.

Musil, C.F. (1993). Effect of invasive Australian Acacias on the regeneration, growth and nutrient chemistry of South African lowland fynbos. *Journal of Applied Ecology*, 30, 361-372.

Ogle, S.M.; Reiners, W. & Gerow, K.G. (2002). Impacts of exotic annual brome grasses (*Bromus spp.*) on ecosystem properties of northern mixed grass prairie. American Midland Naturalist, 149, 46-58.

Oliver, I.; Beattie, A.J. (1993). A possible method for the rapid assessment of biodiversity. *Conservation Biology,* 7, 562-567.

Pyšek, P. & Pyšek, A. (1995). Invasion by *Heracleum mantegazzianum* in differents habitats in Czech Republic. *Journal of Vegetation Science*, 6, 771-718.

Ridley, A.M.; White, R.E.; Helyar, K.R.; Morrison, G.R.; Heng, L.K. & Fisher, R. (2001). Nitrate leaching loss under annual and perennial pastures with and without lime on a duplex (texture contrast) soil in humid southeastern Australia. *European Journal of Soil Science*, 52, 237-252.

Rothstein, D.E.; Vitousek, P.M. & Simmons, B.L. (2004). An exotic tree alters decomposition and nutrient cycling in a Hawaiian Montane Forest. *Ecosystems*, 7, 805-814.

Scheffer, R.A. & Aerts, R. (2000). Root decomposition and soil nutrient and carbon cycling in two temperate fen ecosystems. *Oikos*, 91, 541-549.

Scott, N.A.; Saggar, S. & McIntosh, P.D. (2001). Biogeochemical impact of *Hieracium* invasion in New Zealand's grazed tussock grasslands: sustainability implications. *Ecological Application*, 11, 1311-1322.

Standish, R.J.; Williams, P.A.; Robertson, A.W.; Scott, N.A. & Hedderley, D.I. (2004). Invasion by a perennial herb increases decomposition rate and alters nutrient availability in warm temperate lowland forest remnants. *Biological Invasions*, 6, 71-81.

Tamm, C.O.; Aronsson, A. & Popovic, B. (1995). Nitrogen saturation in a long-term forest experiment with annual additions of nitrogen. *Water, Air & Soil Pollution*, 85, 1683-1688.

Topp, W.; Kappes, H. & Rogers, F. (2008). Response of ground-dwelling beetle (*Coleoptera*) assemblages to giant knotweed (*Reynoutria spp.*) invasion. *Biological Invasions*, 10, 381-390.

Vanderhoeven, S.; Dassonville, N. & Meerts, P. (2005). Increased topsoil mineral nutrient concentrations under exotic invasive plants in Belgium. *Plant and Soil*, 275, 167-177.

Wardle, D.A.; Nicholson, K.S. & Rahman, A. (1995). Ecological effects of the invasive weed species *Senecio jacobea* L. (ragwort) in a New Zealand pasture. *Agriculture, Ecosystems & Environment,* 56, 19-28.

Wedin, D.A. & Tilman, D (1996). Influence of nitrogen loading and species composition on the carbon balance of grasslands. *Science*, 274, 1720-1723.

Whitehead, D.C. (1995). *Grassland Nitrogen*. CAB International, Wallingford, UK.

Wilsey, B.J.; Polley, H.W. (2006). Aboveground productivity and root–shoot allocation differ between native and introduced grass species. *Oecologia*, 150, 300-309.

In: Invasive Species: Detection, Impact and Control
Editors: C.P. Wilcox and R.B. Turpin ISBN 978-1-60692-252-1

Chapter 6

EFFECTS OF NITROGEN FORM ON THE GROWTH AND PHYSIOLOGICAL RESPONSES OF AN INVASIVE AQUATIC PLANT *EICHHORNIA CRASSIPES*

***Weiguo Li*[a, b] *and Jianbo Wang*[a *]**

[a]Key Laboratory of the MOE for Plant Developmental Biology, College of Life Sciences, Wuhan University, Wuhan, 430072, China

[b]Institute of Resource and Environment, Henan Polytechnic University, Jiaozuo, 454003, China

ABSTRACT

Increased input of nitrate and ammonium to ecosystems is mainly responsible for eutrophication, which is related to biological invasion of aquatic exotic plant species. In order to determine the effects of different nitrogen forms on its growth and physiological responses in eutrophic water, *Eichhornia crassipes* plants were grown for 28 days in 5mmol/L nitrogen-contained nutrient solutions varying in NH_4^+:NO_3^- ratio (100:0, 75:25, 50:50, 25:75, 0:100) in laboratory. The results showed that the NH_4^+:NO_3^- ratio of nutrient solution dramatically affected the plant performance of *E. crassipes* including relative growth rate, number of generated ramets, nitrate concentration in root and leaf. Furthermore, nitrate reductase activity increased with reduction of NH_4^+: NO_3^- ratio in culture solution demonstrating preferences of N sources as nitrate in *E. crassipes*. However, ammonium concentration and glutamine synthetase activity in leaf did not significantly change, and those in root significantly increased when proportion of NH_4^+ in nutrient solution increased indicating that *E. crassipes* plants are able to resist ammonium toxicity by regulation of ammonium transport and assimilation. In a word, the results suggested that establishment and expansion of *E. crassipes* was not likely to be limited by nitrogen forms.

Key words: ammonium tolerance, *Eichhornia crassipes,* NH_4^+-N, NO_3^--N, physiological response

INTRODUCTION

During the last five decades, human activities have substantially altered the global nitrogen cycle, increasing both the availability and the mobility of nitrogen, especially inorganic nitrogen, over large regions of aquatic ecosystems (Vitousek et al., 1997; Carpenter et al., 1998). As a consequence, increased input of inorganic nitrogen has led to eutrophication of many water bodies, and has been attributed to change of ecological process in fresh water ecosystems. One of the significant changes of fresh water ecosystems is biological invasion of exotic aquatic weeds, which bring many serious problems, such as disruption of ecosystems, competition with native species and economic losses (Pimentel et al., 2000). In many cases, it has been shown that invasions in aquatic ecosystems are associated with elevated or fluctuating resource levels, and invasion success is thought to be the combined outcome of nutrient conditions and plant traits in aquatic habitats (Davis and Pelsor, 2001; Zedler and Kercher, 2004; Hastwell et al., 2008). Nitrate and ammonium, two inorganic nitrogen forms, are main components of eutrophic water bodies, and serve as available nitrogen resources for aquatic plant growth. Invasive species often have higher growth rates than that of non-invasive species, especially at high nitrogrn levels. Therefore, invasion of aquatic plant probably attributed to its strong capability to assimilate nitrogen at a fast rate and hence have greater ability to take advantage of high nitrogen availability.

Nitrate is the major nitrogen form for higher plant, and most plant species tolerate high nitrate concentrations without any sign of toxicity. However, ammonium as the sole nitrogen source leads to physiological disorders and reduced growth when compared with nitrate or mixed nitrogen nutrition. Most vascular plants acquire nitrogen as nitrate or ammonium, and the physiological response to different nitrogen forms were reported in many previous studies (Majerowicz and Kerbauy, 2002; Munzarova et al. 2006; Fernandes and Gelvan, 2007). Preference for nitrate or ammonium varies according to the plant species, which is probably related to the physiological adaptations of plants to natural ecosystems (Adams and Attiwill, 1982). In inorganic nitrogen assimilation process, nitrate reducrase and glutamine synthetase are two key enzymes, which regulate nitrogen utilization in plants. It is well established that nitrogen forms displayed seasonal variation, and physiological response to nitrogen forms may be related to ecological performance of plant species (Smirnoff and Stewart, 1986; Paulissen et al, 2004).

Eichhornia crassipes, a fast clonal growth plant, is considered as one of the worst invasive weeds in water body. After establishment, the invader rapidly spreads by generation of the ramets and occupies broad range of habitats in a short time (Howard and Harley 1998). The growth rate of *E. crassipes* seems to exceed that of any other aquatic plants, and this high propagation rate may depend on the peculiar physiological characteristics and nutrient absorption efficiency of water hyacinth. Many studies have dealt with the influence of nutrient supply on phenology, growth, nutrient absorption and storage of *E. crassipes* (Center and Spencer, 1981; Reddy et al., 1989; Greenway, 1997). Although large numbers of data, including biomass production, population dynamics, and phenotypic plasticity showed that advantage of nitrogen utilize in *E. crassipes*, only few investigation focused on its performances of nitrogen assimilation. *E. crassipes* displays more aggressive in adequate either nitrate or ammonium environments, which probably attributed to its flexible capability to assimilate different nitrogen forms effectively. Therefore, the study of the activity of key nitrogen assimilating enzymes may give valuable information on the physiological adaptive mechanism of invasive plant species under different nitrogen sources conditions.

The objectives of the present study are to examine the effects of nitrogen form on: (i) the relative growth rate and clonal growth; (ii) concentration of nitrate and free ammonium in tissues; (iii) the activity of nitrate reductase (NR), glutamine synthetase (GS) in roots and leaves of invasive plant species *E. crassipes*.

MATERIALS AND METHOD

Plant Material

The experimental materials were collected from a eutrophic lake (Nanhu Lake, E 114°21.198, N30°33.118) in Wuhan, China. After removing the mud and epiphyte by washing with 0.5% sodium hypochlorite solution, a few ramets developing from the single clone were selected for pre-incubation, and the growth medium was adapted from one-fourth of Hoagland nutritious solution. Plant materials propagated vegetatively through a few generations in the green house of Wuhan University, to avoid the ramets from different clone. Furthermore, extremely low genetic variation of *E. crassipes* in China was revealed in our previous study (Li et al., 2006), and thus these plants developed from single clone in the present study might not differ in genotype.

Plants were kept in the culture room and were received to a 12/12 h photoperiod. The photon flux measured was 200μmol m^{-2} $s^{-1}$and the temperature was 30/22 °C (day/night) at plantlet level. In order to avoid toxicity of ammonium, an experiment was conducted to detect the effect of ammonium on plant growth. Modified Hoagland solutions, one-fourth strength, were used in order to obtain several concentrations (0.5, 1.0, 3.0, 5.0 and 7.0 mM) of ammonium. The pH of the nutrient solution was kept at 5.8, and was adjusted daily using KOH. No significantly synmpose of plant was observed in each ammonium treatment, and 5.0 mM total nitrogen was selected in further experiments.

Experiment Design

New-generated *E. crassipes* single clone plants (4-5 leaves) with root length not beyond 1 cm were selected to be incubated for our experiment in greenhouse (Wuhan University). The incubation ramets were transferred to full nutrient solution (2 mM KH_2PO_4, 0.5 mM $MgSO_4$, 50 μM FeNaEDTA, 25 μM H_3BO_3, 2 μM $MnSO_4{\cdot}H_2O$, 1 μM $ZnSO_4{\cdot}7H_2O$, 0.5 μM $CuSO_4{\cdot}5H_2O$, 0.2 μM $NaMoO_4$) containing 5mM nitrogen source. Nitrogen was supplied with $NaNO_3$, NH_4Cl, or $NaNO_3$ plus NH_4Cl. Plantlets were grown in the presence of five different $NO3^-$: $NH4^+$ratios: 100:0, 75:25, 50:50, 25:75 and 0:100. The nutrient solution pH value was kept at 5.8, and the growth medium solutions were changed every two days. Each treatment consisted of four experimental replicates.

Harvest

Plants were harvested at 28 days after transfer of single clone plants to grow medium. Numbers of generated clones were recorded in each treatment. Harvested plants were separated into leaves and roots respectively, and four individuals from each treatment were dried at 85 ℃ for 24 h to measure dry weight and nitrate accumulation. At the same time, other fresh roots and leaves under different growth conditions were used to determine tissue ammonium concentration, NR activity, and GS activity.

Tissue Nitrate and Ammonium Determination

Nitrate content was determined with water extraction of dry tissues. Sample of 100 mg dry tissue was dispersed in double-distilled water at 45 ℃ for 60 min. The suspension was centrifuged at 5000g for 15 min, and the nitrate level was measured in the supernatant using salicylic-acid method according to Cataldo et al. (1975). Absorbance was measured at 410 nm. Tissue ammonium concentration was determined by the indophenol-blue reaction (Solozano, 1969). Fresh plant material was grounded with liquid nitrogen, dissolved in distilled water (1 g fresh weight (FW) to 10 ml of distilled water), and extracts were heated at 70℃ for exactly 2 min, centrifuged at 5,000 g for 5 min at 4℃ and ammonium determined in the supernatant. The amount of ammonium was expressed as μmol NH_4^+ g^{-1} FW.

Enzyme Activity Measurement

In vitro NR activity assays were performed according to a modification of Kaiser and Huber (1997). Plant material (500 mg) were grounded to fine powder with liquid nitrogen, and suspended in 1.5 ml of 50 mM HEPES-KOH buffer (pH 7.5), containing 10 μM FAD, 5 mM $MgCl_2$, 0.5 mM EDTA, 5 mM DTT, 0.1% Triton X-100 (w/v), 10% glycerol (v/v), 50μM leupeptin, 1% PVPP (w/v), 0.2 mM PMSF. The homogenate was centrifuged for 20 min at 1, 5000 g and 0.25 ml of supernatant was added with 0.75 ml assay mixture consisting of 50 mM HEPES-KOH buffer (pH 7.5), 10 mM KNO_3, 0.2 mM NADH, and 2 mM EDTA for NRA and incubated at 30℃ for 10 min. The reaction was stopped by adding 0.05 ml 0.5M Zn acetate. Excess NADH was destroyed with 10 mM phenazine methosulphate. The amount of nitrite released in 0.5 ml reaction mixture was estimated by adding 1 ml of 1% sulfanilamide in 3 M HCl (w/v) and 1 ml of 0.02% *N*- (1-naphthyl) ethylene diamine dichorite solution, and the absorbance was read at 540 nm. The amount of nitrite was read from a standard curve of nitrite. NR activity is expressed as μmol NO_2^- g^{-1} FW h^{-1}.

Total GS activity was assayed following the modified method of Rhodes et al. (1978). Fresh tissue (500 mg) was grounded in liquid nitrogen and suspended with 1.5 ml of Tris-HCl buffer (pH 7.6) containing 10 mM 2-mercaptoethanol, 1 mM EDTA, 1 mM $MgCl_2$, 0.5 mM PMSF, 50μM leupeptin, and 10% glycerol (v/v). The homogenates were centrifuged at 15000 g for 30 min. The assay mixture consisted of 100 mM imidazole buffer (pH 7.2), 11.4 mM ATP, 45 mM $NH_2OH\cdot HCl$, 45 mM $MgSO_4$ and the enzyme extract. After incubation at 37 ℃ for 15 min, and the reaction was terminated by adding $FeCl_3$ reagent (10% $FeCl_3$ in 0.2 M

HCl, 24% TCA and 50% (v/v) HCl mixed in the ratio of 1:1:1). Absorbance of the supernatant was measured at 540 nm after centrifugation at 6000 g for 5 min. One unit of enzymatic activity was defined as the formation of μmol γ-glutamyl hydroxamate per hour.

Calculations and Statistics

The relative growth rate (RGR) of *E. crassipes* based on net dry weight increment was calculated according to Hunt (1982):

$$RGR = (\log_e W_2 - \log_e W_1)\, t^{-1}$$

where W_1 and W_2 are initial and finial sample dry weight and t is culture time.

Data were statistically analyzed using the one-way analysis of variance. Tukey's multiple range test was used to compare the means of all nitrogen treatments. Statistical analyses of the data were carried out at $P < 0.05$ significance level using SPSS software package (release 11.5, SPSS Inc., Chicago, IL, USA).

RESULT

RGR and Clone Growth

Considered the toxically effect of high concentration ammonium on plant, the effect of ammonium on plant growth was detected in order to set up a specific total nitrogen concentration in this study. In terms of relative growth rate, our results showed that plant responses varied between treatments, depending on supplied ammonium concentration (Fig. 1A). The optimal plant growth (0.072 ± 0.007 $gg^{-1}d^{-1}$) was found in plants supplied with 5 mM ammonium, and RGR of *E. crassipes* increased for ammonium concentrations ranging from 0.5 mM to 5mM. RGR (0.069 ± 0.005 $gg^{-1}d^{-1}$) slightly decreased by 4.2% when ammonium concentrations increased from 5mM to 7mM, but there was no significant difference. According to the above results, specific total nitrogen concentration of 5 mM was selected for further study.

After 28 days of incubation in different nitrogen sources, plants supplied with sole nitrate showed better vegetative appearance than those grown in the other $NO_3^-:NH_4^+$ ratios. The highest RGR was 0.094±0.006 $gg^{-1}d^{-1}$, and *E. crassipes* RGR reduced with decrease of partition of nitrate in nutrition solution. Ammonium as the only nitrogen source resulted in relatively low growth rates (0.071±0.005 $gg^{-1}d^{-1}$) when compared to plants grown with other $NO_3^-:NH_4^+$ ratios (Fig. 1A). Similar nitrogen form effect was observed in clone numbers (Fig. 1B). The clone number slightly increased with increase of partition of nitrate in solution, and the highest clone number was 6.75±0.96 in sole nitrate-supplied plants. Except for nitrate-fed plants, there was no significant difference of clone number among different $NO_3^-:NH_4^+$ treatments.

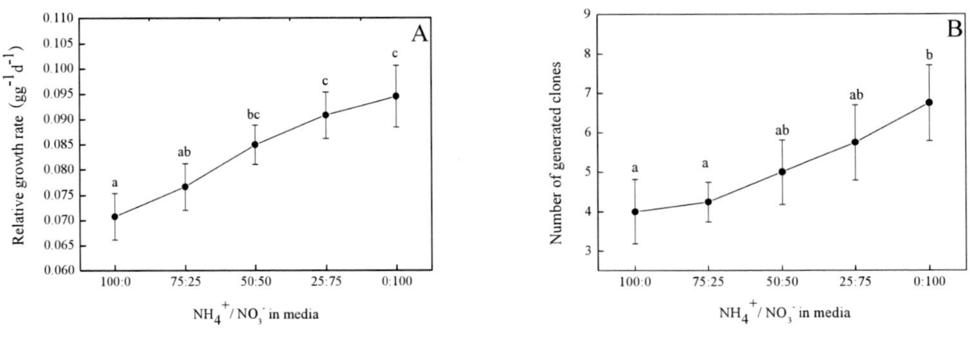

Figure 1. Effects of mixed nitrogen regimes on growth (A) and generated clone number (B) of *E. crassipes*. 5 mM of nitrogen was supplied to plants composed with different ratios of $NaNO_3$and NH_4Cl. Values are the means of four replicates ± SE. Different letters indicate significant differences among treatments at $P < 0.05$ (one-way ANOVA's Tukey test).

Nitrate and Ammonium Concentration

The nitrate concentration was always higher in leaves than that in roots regardless nitrogen form, and it significantly increased in both parts as the nitrate concentration increased in the medium (Fig. 2A). When nitrate was the sole nitrogen source, the highest nitrate concentration in root and leaf were found, which were 6.7 and 7.1 folds higher than that in ammonium-fed plants, respectively. Similarly, higher ammonium concentration was observed in leaves. The free ammonium amounts in both shoot and root tissues were relatively low in nitrate treated plants, and it increased in both parts with increase of ammonium partition in the medium (Fig. 2B). However, changes of ammonium concentration in roots and leaves showed different pattern. Roots ammonium concentration significantly increased with increase of partition of ammonium in solution. The highest ammonium concentration was observed in ammonium-fed plants (1.94±0.14μmol g^{-1}FW), which was four folds than that in nitrate-fed plants. For ammonium content in leaves, it only significantly increased with partition of ammonium when relatively high nitrate form was available in nutrient solution. Under high ammonium condition, there were no significant differences between treatments. For example, the highest free ammonium contents of leaf tissues (2.20±0.15μmol g^{-1}FW) were found in plants grown in the presence of ammonium (NO_3^-/NH_4^+ 0:100). The ammonium contents of leaves were not significantly different and increased as ammonium was supplied in greater proportion ($NH_4^+ /NO_3^- > 1$).

Nitrate Reductase Activity and Glutamine Synthetase Activity

The youngest fully expanded leaves and new-formed roots were used to detect NRA in *E. crassipes*. NRA in roots and leaves were given in Fig. 3A. The results showed that root and leaf NRA in *E. crassipes* were significantly affected by different nitrogen sources, and increased with nitrate partition in media solution. In all NH_4^+ /NO_3^- proportions treatments, the activity of NR showed the same pattern of distribution among plant parts, and high proportion of NRA of *E. crassipes* was observed in leaves. Opposite to NR activity, GS activity in both roots and leaves decreased with increase of nitrate partition in the media (Fig. 3B). The highest rates of GS activity were found in the leaves of *E. crassipes* plants supplied with sole ammonium. NH_4^+ /NO_3^- significantly affected leaf GS activity in plants grown in the media containing NO_3^- as the major nitrogen form. When partition of ammonium in the media increased, leaf GS activity was only slightly influenced by NH_4^+ /NO_3^- ratio. For GS activity in *E. crassipes* roots, there were significantly differences among various treatments.

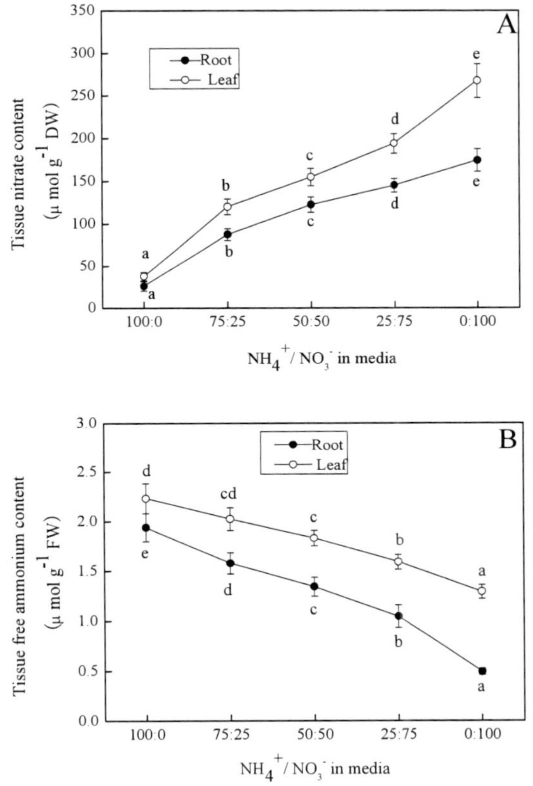

Figure 2. Effects of mixed nitrogen regimes on tissue nitrate concentration (A) and free ammonium content (B) in root (black) and leaf (white). 5 mM of nitrogen was supplied to plants composed with different ratios of $NaNO_3$and NH_4Cl. Values are the means of four replicates ± SE. Different letters indicate significant differences among treatments at $P < 0.05$ (one-way ANOVA's Tukey test).

DISCUSSION

Nitrate and ammonium are the most important inorganic N sources for plants, but they induce different growth effects in most plants studied (Britto and Kronzucker, 2002). Results of this study clearly demonstrated invasive plant *E. crassipes* was able to utilize both nitrate and ammonium as nitrogen sources effectively. However, plants showed different responses in plant performance including relative growth rate and clone number, after four weeks of growth in different NH_4^+:NO_3^- sources. This differences indicates that distinct N sources modulate the growth rates of *E. crassipes*, with NO_3^- being the most efficient source and NH_4^+ the least efficient. NO_3^- gave even better growth results than NH_4^+:NO_3^- and NH_4^+ alone. It is well established that various plant preferences of N sources are one of the nutritional aspects affecting plant community establishment at sites where the NH_4^+/NO_3^- substrate ratio changes during succession (Kronzucker et al., 1997). While some studies reported plant prefers for NO_3^- over NH_4^+ in wetland plant, and others reported the opposite effect in submerge macrophytes (Tylová-Munzarová et al, 2005; Tylová et al, 2008). For invasive plant *E. crassipes*, it floats on the surface of disturbed water body but showed different preference from submerge macrophytes. Previous study demonstrated again that different plant species require different forms of N, and plant preference for different forms of N is influenced by environmental factors such as root- or air-temperature and solution pH. Both pH and temperature have been shown to affect N uptake rates and differentially affect ammonium and nitrate uptake rates in some cases (Garnett and Smethurst, 1999). For example, low root temperature generally reduce uptake rate, but may reduce nitrate uptake to a greater extent than ammonium uptake (Macduff and Jackson, 1991). As a tropic origin plant species, it may be an adaptation of *E. crassipes* to grow in open water columns where nitrate is the predominant N source. On the other hand, we chose pH 5.8 as being representative of acidic water columns. In general, high value of pH depressed ammonium uptake than nitrate uptake, and the preference of nitrate *in E. crassipes* may also result from the relatively high value of pH in nutrient media in present study.

Significant amounts of NO_3^- usually occur in the bulk water, top sediment layers in wetland, and nitrate as nitrogen resource is available for floating macrophytes. Our previous study showed that growth and tissue NR activity of *E. crassipes* was significantly stimulated by increase of nitrate availability (Li and Wang, 2007). In accordance with this, significant increases of the RGR and clone number with nitrate availability in nutrient solution were found in present study, and *E. crassipes* proved as a leaf nitrate reducer. On the other hand, capability of NH_4^+ utilization is documented in wetland plants, and NH_4^+ preference is often suggested as the feature of plants adapted to habitats in which NH_4^+ is prevalent (Cedergreen and Madsen, 2003). However, high ammonium, especially sole ammonium as nitrogen resource, results in potential toxicity to vascular plants, and this NH_4^+ toxicity are often correlated with increased concentrations of NH_4^+ in plant tissues. Although the free ammonium amounts in both shoot and root tissues of *E. crassipes* increased with partition of ammonium in our study, the significant increase was observed only in root especially in high NH_4^+/NO_3^- conditions. Schoerring and his coworkers (2002) reported that one of the main factors responsible for the difference observed in the sensitivity and tolerance of plants to ammonium is related to the ammonium concentration present in the root. The retention of ammonium in the roots of *E. crassipes* might prevent transport of ammonium to the shoot, which is more sensitive to ammonium accumulation. Results of NH_4^+ accumulation in roots under high NH_4^+/NO_3^- condition reported here suggested that *E. crassipes* is tolerant to high ammonium stress.

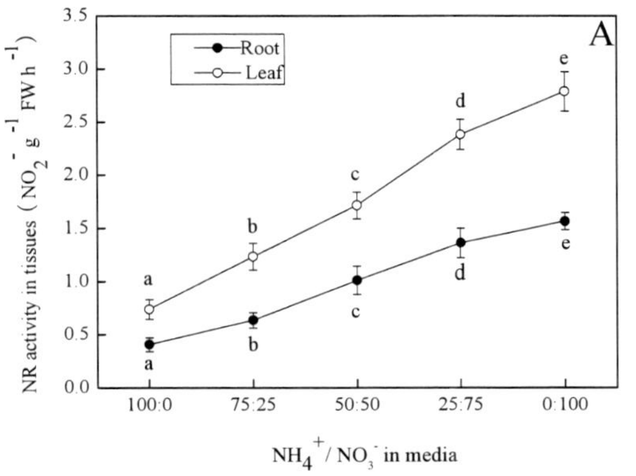

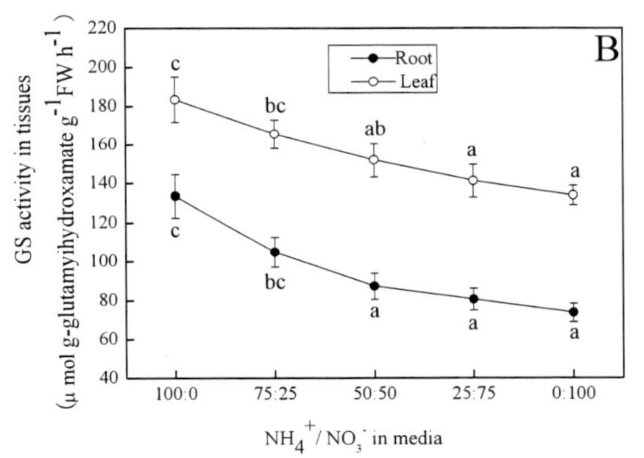

Figure 3. Effects of mixed nitrogen regimes on nitrate reductase activity (A) and glutamine synthetase activity (B) in root (black) and leaf (white). 5 mM of nitrogen was supplied to plants composed with different ratios of $NaNO_3$and NH_4Cl. Values are the means of four replicates ± SE. Different letters indicate significant differences among treatments at $P < 0.05$ (one-way ANOVA's Tukey test).

NR and GS activity was detected in leaves and roots of *E. crassipes* under different NH_4^+/NO_3^- conditions. Since NR is a kind of substrate-induced enzyme, the end products of nitrate assimilation such as ammonium usually inhibit NR activity (Sivasankar and Oaks, 1996). With the increase of ammonium in media, decreases of NRA in roots and leaves of *E. crassipes* were likely attributed to this inhibited effect of ammonium in our study. Conversely, increase of GS activity in tissues with increase of NH_4^+/NO_3^- ratio. Many studies reported that the influence of nitrogen form on GS activity may be specie-specific in plant. For instance, Kant et al. (2007) reported that GS activity increased in barley when NH_4^+ instead of NO_3^- was used as a nitrogen source. In ryegrass, however, nitrogen ionic form had no effect on GS activity (Sagi et al., 1998). Our results indicated that the mixed nitrogen regime led to increased GS activity compared to NO_3^- alone in *E. crassipes*. It has been suggested that the GS could work as an ammonium stress-adaptive mechanism and play important role in alleviation of ammonium toxicity by incorporation of ammonium to amino acids (Given, 1978). However, relatively high NH_4^+ concentration in leaves was observed in present study in spite of nitrogen forms. Although it has been thought that, mainly under NH_4^+ nutrition conditions, NH_4^+ had to be preferably assimilated in roots because of its possible toxic effects (Tobin and Yamaya, 2001). Significant quantities of NH_4^+ in xylem and apoplastic sap in NO_3^--fed as well as in NH_4^+ -fed plants was reported in previous study (Schjoerring et al. 2002). A possible explanation for higher free NH_4^+ content in leaves of *E. crassipes* is that this situation probably originated from metabolic processes in the leaves and xylem translocation. If root and leaf GS activity of *E. crassipes* keeps at a high level under high ammonium condition, accumulation of ammonium in tissue, which might lead to toxicity, was likely prevented. In general, the works on ammonium assimilation by distinct plant species show that plants with high GS activities are more tolerant to ammonium (Glevarec et al. 2004). Recently, Cruz et al (2006) reported that GS activity was higher in ammonium-tolerant than that in ammonium-sensitive plant species, resulting in that the amount of ammonium translocated from the roots to the shoots and the production of high ammonium concentrations in the shoots would thus decrease. Therefore, our results suggested that the main location of nitrogen metabolism (shoots or roots) and the high levels of GS activity are an important strategy for ammonium tolerance of *E. crassipes*.

CONCLUSION

The results from the experiments indicated that invasive plant *E. crassipes* has capability for utilizing both nitrate and ammonium for fast growth, of which nitrate is preferred as nitrogen sources. Furthermore, resistance to ammonium toxicity by regulation of ammonium transport and assimilation was revealed in *E. crassipes* under high ammonium conditions. The effects of nitrate and ammonium nitrogen on growth, and activity of key enzymes of nitrogen metabolism in *E. crassipes* appear to reflect its different adaptability to water conditions which might have evolved as an important strategy of survival of the invasive species in natural environment. *E. crassipes* which is able to colonize a wide range of water body after a recent natural disturbance in a fresh water ecosystem, must be highly adaptable to different nutrient conditions such as N form. Our study reported here suggested that establishment and expansion of this invasive plant investigated was not likely to be limited by nitrogen forms.

ACKNOWLEDGEMENTS

The authors thank Miss Jingjing Shen for her technical assistance. This work was carried out with the financial support from the National Natural Science Foundation of China (No. 30521004, 30170177) and Basic and Frontier Research Program from Henan Province (No. 082300430350).

REFERENCES

Adams, M.A., Attiwill P.M. (1982). Nitrate reductase activity and growth response of forest species to ammonium and nitrate sources of nitrogen. *Plant and Soil, 66*, 371-383.

Britto, D.T., Kronzucker, H.J. (2002). NH_4^+ toxicity in higher plants: a critical review. *Journal of Plant Physiology, 159*, 567-584.

Carpenter, S.R., Caraco N.F., Correll, D.L., Howarth, R.W., Sharpley, A.N., Smith, V.H. (1998). Nonpoint pollution of surface water with phosphors and nitrogen. *Ecological Applications*, *8*, 559-568.

Cataldo D.A., Haroon M., Schrader L.E., Youngs V.L. (1975). Rapid calorimetric determination of nitrate in plant tissues by nitration of salicylic acid. *Communication of Soil Science and Plant Analysis, 6*, 71-80.

Cedergreenand, N., Madsen T.V. (2003). Nitrate reductase activity in roots and shoots of aquatic macrophytes. *Aquatic Botany, 76*, 203-212.

Center, T.D., Spencer, N.R. (1981). The phenology and growth of water hyacinth (*Eichhornia crassipes* [Mart.] Solms) in a eutrophic north-central Florida lake. *Aquatic Botany*, *10*, 1–32

Cruz, C., Bio, A.F.M., Domínguez-Valdivia, M.D., Aparicio-Tejo, P.M.□Lamsfus, A., Martins-Loucão, C, M. (2006). How does glutamine synthetase activity determine plant tolerance to ammonium? *Planta, 223*: 1068-1080.

Davis, M.A., Pelsor, M. (2001). Experimental support for a resource-based mechanistic model of invasibility. *Ecology Letters, 4*, 421–428.

Fernandes, E., Gelvan, A. (2007). Inorganic nitrogen assimilation *Chlamydomonas*. *Journal of Experimental Botany, 58*, 2279-2287.

Garnett, T.P., Smethurst, P. J. (1999). Ammonium and nitrate uptake by *Eucalyptus nitens*: effects of pH and temperature. *Plant and Soil, 214*, 133–140.

Given, C.V. (1978). Metabolic detoxification of ammonium in tissues of higher plants. *Phytochemistry, 18*, 375–382.

Glevarec, G., Bouton, S., Jaspard, E., Riou, M-T., Cliquet, J-B., Suzuki,A., Limami, A.M. (2004). Respective roles of glutamine synthetase/ glutamate synthase cycle and glutamate dhydrogenase in ammonium and amino acid metabolism during germination and post-germinative growth in the model legume *Medicago truncatula*. *Planta, 219*, 286–297.

Greenway, M. (1997). Nutrient content of wetland plants in construct wetlands receiving municipal effluent tropic Australia. *Water Science and Technology, 34*, 407-412.

Hastwell, G.T., Daniel, A.J., Vivian-Smith G. (2008). Predicting invasiveness in exotic species: do subtropical native and invasive exotic aquatic plants differ in their growth responses to macronutrients? *Diversity and Distributions, 14,* 243–251.

Howard, G.W., Harley, K.L.S. (1997). How do floating aquatic weeds affect wetland conservation and development? How can these effects be minimized? *Wetland Ecology and Management*, *5*, 215-225.

Hunt, R. (1982). *Plant Growth Curves: The Functional Approach to Plant Growth Analysis.* London, Edward Arnold Press.

Kaiser, W.M., Huber, S.C. (1997). Correlation between apparent activation state of nitrate reductase (NR), NR hysteresis and degradation of NR protein. *Journal of Experimental Botany, 48,* 1367-1374.

Kronzucker, H.J., Siddiqi, M.Y., Glass, A.D.M. (1997). Conifer root discrimination against soil nitrate and the ecology of forest succession. *Nature, 385*, 59–61.

Kant, S., Kant, P., Lips, H., Barak, S. (2007). Partial substitution of NO_3^- by NH_4^+ fertilization increases ammonium assimilating enzyme activities and reduces the deleterious effects of salinity on the growth of barley. *Journal of Plant Physiology, 164*, 303-311.

Li, W.G., Wang, B.R., Wang, J.B. (2006). Lack of genetic variation of an invasive clonal plant *Eichhornia crassipes* in China revealed by RAPD and ISSR markers. *Aquatic Botany, 84*, 176-180.

Li, W.G., Wang, J.B. (2007). Effects of light and nitrate on growth and physiological characteristics of invasive plant *Eichhornia crassipes* (Mart.) Solms. – *Journal of Wuhan University (Natural Science Edit), 53*, 457-462.

Macduff, J.H., Jackson, S.B. (1991). Growth and preferences for ammonium and nitrate uptake by barley in relation to root temperature. *Journal of Experimental Botany, 42*, 521-530.

Majerowicz, N., Kerbauy, G. B. (2002). Effects of nitrogen forms on dry matter partitioning and nitrogen metabolism in two contrasting genotypes of *Catasetum fimbriatum* (Orchidaceae). *Environmental and Experimental Botany, 47*, 249-258.

Munzarova, E., Lorenzen, B., Brix, H., Vojtiskova, L., Votrubova, O. (2006). Effect of NH_4^+/NO_3^- availability on nitrate reductase activity and nitrogen accumulation in wetland helophytes *Phragmites australis* and *Glyceria maxima*. *Environmental and Experimental Botony, 55*, 49-60.

Paulissen, M.P.C.P., van der Ven, P., Dees, A, Bobbink, R. (2004). Differential effects of nitrate and ammonium on three fen bryophyte species in relation to pollutant nitrogen input. *New Phytologist, 164*, 451–458.

Pimentel, D., Lach, L., Zuniga, R., Morrison, D. (2000). Environmental and economic costs of nonindigenous species in Unite States. *Bioscience, 50*, 53-65.

Reddy, K.R., Agami, M., Tucker, J.C. (1989). Influence of nitrogen supply rates on growth and nutrient storage by water hyacinth (*Eichhornia crassipes* (Mart.) Solms) plants. *Aquatic Botany, 36*, 33-43.

Rhodes, D., Rendo, G.A., Stewart, G.R. (1975). The control of glutamine synthetase level in *Lemna minor* L. *Planta, 125*, 201-211

Sagi, M., Dovrat, A., Kipnis, T., Lips, S.H. (1998). Nitrate reductase, phosphoenolpyruvate carboxylase and glutamine synthetase in annual ryegrass as affected by salinity and nitrogen. *Journal of Plant Nutrition, 21*, 707–723.

Schjoerring, J.K., Husted, S., Mack, G., Mattsson, M. (2002). The regulation of ammonium translocation in plants. *Journal of Experimental Botany, 53*, 883–890.

Sivasankar, S., Oaks, A. (1996). Nitrate assimilation in higher plants: The effect of metabolites and light. *Plant Physiology and Biochemistry, 34*, 609–620.

Smirnoff, N., Stewart, G.R. (1986). Nitrate assimilation and translocation by higher plants: comparative physiology and ecological consequences. *Physiologia Plantarum, 64*, 133–140

Solozano, L. (1969). Determination of ammonium in natural waters by the phenolhypochlorite method. *Limnologic Oceanography, 14*, 799-801.

Tobin, A.K., Yamaya, T. (2001). Cellular compartmentation of ammonium assimilation in rice and barley. *Journal of Experimental Botany, 52*, 591-604.

Tylová, E., Steinbachová, L., Votrubová, O., Lorenzen, B., Brix H. (2008). Different sensitivity of *Phragmites australis* and *Glyceria maxima* to high availability of ammonium-N. *Aquatic Botany, 88*, 93-98.

Tylová-Munzarová, E., Lorenzen, B., Brix, H., Votrubova′, O. (2005). The effects of NH_4^+ and NO_3^- on growth, resource allocation and nitrogen uptake kinetics of *Phragmites australis* and *Glyceria maxima*. *Aquatic Botany, 81*, 326–342.

Vitousek, P.M., Aber, J.D., Howarth, R.W., Likens, JE; Maston, PA; Schindler, DW; Schlesinger, WH; Tilman, DG. (1997). Human alteration of the global nitrogen cycle: sources and consequences. *Ecological Applications*, *7*, 737-750.

Zedler, J.B., Kercher, S. (2004). Causes and consequences of invasive plants in wetlands: opportunities, opportunists, and outcomes. *Critical Reviews in Plant Sciences, 23*, 431-452.

In: Invasive Species: Detection, Impact and Control
Editors: C.P. Wilcox and R.B. Turpin
ISBN 978-1-60692-252-1

Chapter 7

INVASIVENESS OF SPECIES USEFUL AS WARM-SEASON PASTURE LEGUMES IN THE SOUTHEASTERN UNITED STATES

W.D. Pitman
Louisiana State University Agricultural Center, Hill Farm Research Station,
Homer, Louisiana 71040, USA

ABSTRACT

Warm-season perennial legumes, as a plant functional group, hold considerable promise for use as forage plants in the warm, humid southeastern U.S., where infertile soils and low-protein forage grasses are common. This plant group is large with tremendous ranges in growth forms and propagation methods. Despite the overall promise, only a few species have been, are currently, or even appear to be potentially useful forage plants. The few species of this group which are widely adapted across the Southeast are also considered, at least by some, to be invasive. Included are a vine (kudzu, *Pueraria montana* var. *lobata*), a tree (mimosa, *Albizia julibrissin*), and a herb (sericea lespedeza, *Lespedeza cuneata*). Rather casual observation can quickly provide evidence of the invasiveness of kudzu. Determination of the invasiveness of sericea lespedeza depends to some extent on the evidence assessed. With mimosa, the question may be about forage value rather than invasiveness. Most of the seemingly unstoppable spread of kudzu is from rapid vegetative extension, while seed provide the means for increase in range of the other two. Simple single-application treatments of various defoliation procedures or even the most effective herbicide have either had minimal or rather temporary effects on kudzu and sericea lespedeza. Long-term strategies, rather than single control treatments, are required for success. Knowledge of basic mechanisms of plant dispersal and ecosystem susceptibility to invasion by sericea lespedeza are needed to allow appropriate decisions about control and prevention of problem populations of this species. Developing technology holds promise for future use of mimosa as an intensively managed forage legume despite invasiveness. The greatest potential forage use of introduced, perennial, warm-season legumes in the southeastern U.S. in general may well be through the use of grazing livestock as part of carefully planned control strategies.

INTRODUCTION

For the past 50 years, relatively inexpensive nitrogen fertilizer has been the primary determinant of forage productivity in the southeastern U.S. As was the case prior to this period, the recent dramatic increases in price of inorganic nitrogen could now be changing the status of pasture legumes from simply desirable to essential for economic levels of productivity. Temperate (cool season) legume species such as clovers (*Trifolium* species) and alfalfa (*Medicago sativa*) are useful forage plants and nitrogen sources in the region, but the long summer growing season also provides tremendous opportunity for effective use of warm-season legume species. The list of warm-season legumes useful as forage plants prior to the extended era of inexpensive nitrogen fertilizer [Sturkie, 1951] is very similar to current lists of available species [as illustrated by Ball et al., 2002]. As noted by Sturkie in 1951, the annual species on this list are excessively expensive due to direct cost and risk associated with annual establishment. The remaining species include only sericea lespedeza and kudzu as perennial warm-season legumes from the early list with rhizoma peanut (*Arachis glabrata*) added in recent years for specific uses in a portion of the region. The additional widely adapted, naturalized woody legume mimosa has been identified as a promising forage plant [Bransby, 1993]. All three of the widely adapted species, sericea lespedeza, kudzu, and mimosa, are introduced species which are on various lists of invasive species in the southeastern region. In consideration of the likely increasing forage demand for these species in the southeastern U.S., some reconciliation of their continuing recommendation for forage use with their identification as invasive species in the region appears warranted. Additional less widely adapted tropical species provide similar situations for specific ecosystems in peninsular Florida. Rather than an exhaustive review of each species, this overview of ecological aspects potentially affecting invasiveness and control of the perennial warm-season legumes useful as forage plants will emphasize recent advances.

KUDZU

Although there is no controversy regarding the invasiveness of kudzu, recognition of the forage value of this species is still important. With the increasing cost of available approaches of kudzu control and the tremendous resilience of established populations, generating some returns from grazing at appropriate seasons and extents to deplete root reserves may allow greater progress in control of a continuing invasion. Currently available approaches to control, and eventually eradicate, this pest are largely based on the expensive repeated application of herbicides, cultivation, mechanical defoliation, grazing, or some combination of these treatments and possibly with facilitation of burning [Miller, 1996; Moorhead and Johnson, 1998; Everest et al., 1999]. Other than the more recent introduction of herbicides, these are the methods which were successfully used to keep kudzu under control on small farms across the southeastern states during the period of initial use of this plant as a shade, erosion control, and forage plant. Small diversified farms included interspersed cultivated fields and grazed pastures which held kudzu in check. As noted by Forseth and Innis [2004] the extensive planting of kudzu followed by widespread consolidation of farmlands and the associated concentration of the southern population in urban areas left countless small, widely distributed kudzu plantings essentially without limitations. Areas not subjected to cultivation or grazing, such as the extensive land areas allowed to develop as secondary forest, were not only readily accessible to any adjacent kudzu plants, but the spreading populations were not monitored and were essentially unnoticed in many instances. As a result, kudzu is now recognized as "perhaps the worst weed problem" in southeastern forests [Demers and Long, 2002]. Although urban and roadside populations were certainly noticed, their control was often neglected.

The need for caution with kudzu plantings was well understood early in the last century as illustrated by the advice of Duggar in 1925 that the plant should not be grown in gardens, along fence rows, on even "near" rich bottomland. The rapid growth and invasiveness of this plant have more recently been widely recognized [Hipps, 1994; Everest et al., 1999; Matlack, 2002]. Potential to drastically alter community structure and displace native vegetation are concerns for natural areas [Smith et al., 2001]. Physiological and ecological aspects of kudzu growth and development supporting this tremendous potential to spread across the landscape have recently been reviewed by Forseth and Innis [2004]. The particular combination of high allocation of photosynthate to extension growth, extensive rooting of vines at nodes, high leaf area indices with efficient light distribution to leaves within the canopy by paraheliotropic leaf movement, ability to fix atmospheric nitrogen, ability to overtop the tallest forest competitors using the structure of trees to reach full sunlight, and ability to exclude most competing plants from a site are some key characteristics contributing to vegetative expansion and dominance by existing populations [Forseth and Innis, 2004]. Tremendous root storage development over a period of years provides resilience under defoliation and, to some extent, resistance to herbicide treatment. Although natural spread by seed is likely very limited [Webster et al., 2006], hardseededness provides opportunity for soil seedbanks to contribute to re-establishment of stands as does root-crown dormancy induced by some effective herbicide treatments [Miller, 1996]. Thus repeated treatment, whether from herbicide, defoliation, or cultivation, and continued monitoring are required for elimination or even simply limitation of existing kudzu populations [Miller, 1996; Moorhead and Johnson, 1998; Everest et al., 1999]. On small scales, application of herbicide directly to remnants of individual plants as spot spray or even direct injection into individual root crowns following an initial defoliation has been recommended [Smith et al., 2001].

Efforts to eradicate kudzu have widely failed primarily due to either lack of appreciation of the continuing effort required and/or failure to recognize the re-colonization potential from population remnants beyond the boundary of the targeted area. The traditional control methods with greater temporal and spatial limitations than the target to be controlled continue to provide disappointing results. An evaluation of an integrated approach in forest establishment indicated that sustained kudzu control from a persistent herbicide could provide sufficient time for a dominant, smothering canopy of loblolly pine at high population density to develop and suppress the herbicide-weakened kudzu population [Harrington et al., 2003]. An appropriate, persistent herbicide with such a selective activity, however, is not available. As is suggested to contribute to natural limitation of this and other invasive plants in their native environments, biological control methods appear to hold promise [Britton et al., 2002]. Considerable progress has been made with evaluation of the fungus *Myrothecuim verrucaria* [Boyette et al., 2002], including development of a patented control process [Boyette et al., 2001]. Two insect species from China have also recently demonstrated potential [Frye et al., 2007]. These novel biological control approaches require considerable additional evaluation and verification of lack of non-target biological effects, particularly to the related and economically important soybean crop. Current populations in the southeastern U.S. are not without pests. Several insects and bacterial pathogens attack the foliage and seed of kudzu, but infestations do not occur at sufficient levels for control of the plant [Webster et al., 2006].

Grazing by domestic livestock represents an additional biological control approach which has proven potential as noted by recommendations for grazing management of kudzu pastures. As stated by Ball et al. [2002], "Continuous close grazing will weaken plants, reduce productivity, and eventually eliminate stands." Miller [1996] noted that "Close grazing for 3 to 4 years can eliminate kudzu when 80 percent or more of the vegetative growth is continuously consumed." Heavy grazing pressure in August and September each year were recognized as particularly effective for control by grazing defoliation [Miller, 1996]. Additional treatments such as dormant-season burning or mechanical methods may be required to remove vines from trees and other structures beyond reach of grazing livestock. Increased demand for goat meat in the region has contributed to interest in using goats for control of kudzu [Bonsi et al., 1992; Luginbuhl et al., 1996] with interest even in urban situations [Jonsson, 2003]. Miller [1996] noted, however, that grazing by cattle had demonstrated the most success at eradication of this plant. Requirements of grazing animals and their control, especially water and fencing, have been substantial limitations to use of this approach. Portable electric fence materials and relatively low-cost water transport equipment contribute to feasibility of this control approach in many situations. Even tethering and hand watering of individual animals may be useful approaches for small plant populations in some urban areas. Recognition of forage value of the plant in addition to grazing as an economical control method could contribute to development of viable grazing services for use and eventual elimination of existing populations in both extensive and localized urban situations.

Additional developments in rather wide ranging fields also have potential to contribute to the control of kudzu. Genetic variation and heterozygosity within populations have been reported at levels to indicate substantial sexual reproduction and selection for heterozygous individuals [Pappert et al., 2000]. Recent assessments of genetic diversity within southeastern U.S. populations indicate multiple introductions and subsequent gene exchange and recombination [Sun et al., 2005]. This information may be particularly useful in understanding responses to control efforts if variable responses among populations are encountered. Monitoring of responses to control efforts is particularly critical to treatment success. Remote sensing technology has shown promise as a useful means of monitoring populations of kudzu [Li et al., 2001].

In addition to developments with potential to aid in control and possibly eventual elimination of kudzu, additional changes now occurring contribute to an increased immediacy, if not actual urgency, for this control. The pathogen causing potentially devastating soybean rust has recently arrived in the region. The ability of this pathogen to over-winter on kudzu creates a substantial and immediate cause for concern [Frye et al., 2007]. In addition, evidence indicates that global warming and increased carbon dioxide concentrations will likely increase competitive ability of kudzu and extend its range to the north in the U.S. [Sasek and Strain, 1988, 1989, 1990; Forseth and Innis, 2004]. These increasing risks from the continuously expanding populations of kudzu are in addition to the already substantial economic losses particularly to forest enterprises, environmental degradation of natural areas, and safety concerns due to obstruction of visibility of traffic and other hazards.

SERICEA LESPEDEZA

Sericea lespedeza presents quite a paradox. It has long been, and continues to be, acclaimed as a highly valuable erosion control and reclamation plant. It has recently become widely recognized as an invasive weedy species, while also being considered an under-utilized forage plant. In contrast to kudzu, seed of sericea lespedeza continues to be commercially marketed and rather widely recommended for continued planting in the southeastern U.S. Agronomically, sericea lespedeza is recognized as difficult to establish and easily lost with excessive defoliation. Ecologically, sericea lespedeza is increasingly recognized as highly invasive and difficult to control. From combined contributions of these rather divergent perspectives, aspects of population dynamics from plant establishment to dominance and eventual population depletion are being discerned. Establishment limitations include hardseededness which adversely affects immediate germination and establishment but enhances opportunity for multiple establishment events beneficial to colonization of unpredictable and unmanaged environments. Initial vegetative growth of improved varieties each growing season is reported to be generally acceptable to grazing livestock and of good nutritive value despite moderate tannin levels. Rapid stem development during the growing season decreases palatability and often leads to the impression that the plant lacks forage value.

Sericea lespedeza is considered a semi-woody, upright-growing herb. Although seedlings are single-stemmed, mature plants can produce numerous upright stems from the perennating plant crown. In pure stands, a dense leaf canopy can develop to shade the soil surface and exclude other plants below this canopy. Senescence of leaves in the lower canopy of undefoliated plants late in the growing season can produce a layer on the soil surface which further excludes other plant species. In undisturbed stands, this layer can build up over a period of years [Baldridge, 1957]. With a canopy height of 1.5 to 2 m, sericea lespedeza is highly competitive with plants below this height, but it is vulnerable to competition from dense-growing plants above this height. Partial shade along forest roads in Louisiana appears to moderate sericea lespedeza with sparse populations existing as non-dominant community components in some instances for decades. Such partial shade from a forest canopy has also been noted to suppress vegetative competition and enhance establishment of sericea lespedeza planted on Louisiana Coastal Plain sites [Pitman, 2006]. Vegetative spread, other than increases in size of plant crowns and number of stems per crown, does not occur to effectively increase populations. Population increase is exclusively from seed, which can exceed 1,000 seeds per stem annually [Ohlenbusch et al., 2001] or several hundred kilograms per hectare of seed under cultivation. Seed are small (approximately 860,000 seed per kilogram), and seedlings are typically slow to develop. They are, therefore, vulnerable to competition, especially from rapidly developing, early colonizing annual grasses in humid environments.

Despite lack of seedling vigor, the ability of sericea lespedeza to spread from test plantings and volunteer in nearby areas was evident from observations of an 1899 planting in Virginia [Henson, 1957; Cope, 1957]. This ability to volunteer has been recognized as a desirable characteristic in pastures of the southeastern U.S. [Ball and Mosjidis, 1995]. Along with the ability to spread under certain conditions, limited palatability or requirement for a period of adjustment to sericea lespedeza by grazing livestock was recognized during early evaluations [Baldridge, 1957]. These indications of potential weediness and resistance to grazing were, however, countered by the establishment limitations and stand loss in heavily defoliated pastures and hay fields. Varieties of sericea lespedeza which were developed for reclamation rather than forage produced stands that were dominant and eventually recognized as inadequate for most economic uses such as livestock production in the coal-mining region of Appalachia. By the early 1980s, research was in progress to develop strategies to convert these dense stands of planted sericea lespedeza to better forage species [Dove et al., 1997]. During this same time period, widespread and increasing occurrence of sericea lespedeza was being reported in Oklahoma and Kansas [Koger and Stritzke, 2004]. The rapid expansion in Kansas was attributed at least in part to large-scale planting of native grass seed contaminated with sericea lespedeza seed as part of the Conservation Reserve Program [Kilgore et al., undated]. In Georgia and north Florida during this same time period, instances of undesired spread of this plant were also noted as illustrated by Landers and Mueller [1986]. From these and other similar divergent reports of the spread of, dominance by, and previously unappreciated limitations of this plant, sericea lespedeza became widely labeled as an undesirable invasive species. Only in the current century, however, have the various aspects of sericea lespedeza distribution, colonization, and dominance been evaluated in terms of ecological and anthropogenic relationships and their interactions.

Brandon et al. [2004] evaluated the mechanisms involved in dominance of an early successional, old field community by sericea lespedeza. They reported that mowing in spring or spring and fall increased sericea lespedeza, while nutrient enrichment (fertilization with nitrogen, phosphorus, and potassium) decreased cover, stem density, and biomass of this plant. The increase in sericea lespedeza with mowing was at least partially due to indirect effects since mowing prevented dominance by a taller growing woody species which shaded and out-competed the sericea lespedeza in un-mowed treatments. Brandon et al. [2004] concluded that light exclusion was the primary mechanism allowing sericea lespedeza to dominate adjacent herbaceous plants and also to be dominated by a dense-growing taller woody plant. The competitive advantage of sericea lespedeza in low nutrient environments was offset by fertilization providing opportunities for species more responsive to higher nutrient levels. Relationships of sericea lespedeza population dynamics to mowing have been further explored [Munger, 2004; Smith et al., 2001] with effects dependent on timing and subsequent treatments. Early spring defoliation, whether by mowing or burning, can remove old vegetation and enhance germination and seedling growth of sericea lespedeza when the soil seedbank includes this species. Repeated mowing or grazing at this time can decrease survival of these seedlings and be used to deplete soil seed reserves. Fall mowing, when seed have matured but not shattered, can effectively disperse the seed with limited effects on the senescent plants. According to Smith et al. [2001], the most effective time to mow for control of sericea lespedeza is during the late bud stage from mid July to late August. Mowing at this stage, particularly for two or three consecutive years, can effectively deplete stored energy because reserves have already been mobilized for reproductive development and will be further depleted to support vegetative regrowth.

The mechanisms of light control by means of a dominant canopy, competitive advantage for limited nutrients, and responsiveness to timing of defoliation or other disturbance providing opportunity for seed dispersal or germination and seedling development appear to be key aspects of developing and maintaining dominance by sericea lespedeza. Successional processes may be arrested indefinitely where a dense canopy of this species develops in resource-limited environments without potential for taller-growing species. In environments where resources can support shrub or forest communities, successional processes may be substantially delayed by a dense population of sericea lespedeza, but disturbance and edge effects will eventually allow increase of woody plants and suppression of sericea lespedeza as the overstory canopy closes. These relationships provide opportunity for sericea lespedeza to exert long-term effects on plant communities. Modifying these relationships may increase the rate of change from dominance by sericea lespedeza to woodlands or divert the trajectory of succession away from a developing dominance by sericea lespedeza. During decades of use in pastures and hay fields, defoliation practices served as effective disturbance mechanisms to control dense populations. Fertilization practices enhanced competitive ability of other pasture plants frequently resulting in depletion of sericea lespedeza stands. Thus, particular opportunity may exist for use of strategic disturbance to allow woody species to establish in appropriate environments. In grassland ecosystems, nutrient enrichment rather than woody plant establishment may be a key component to combine with strategic disturbance to suppress sericea lespedeza. Timing and cost of disturbance events must be appropriate for the problem presented. Fire and grazing, with appropriate timing and frequency, perhaps provide the most cost-effective means of disturbance for established populations. Appropriately timed mowing and herbicide application may also be effectively used [Altom et al., 1992; Jordan et al., 2002; Koger et al., 2002; Cummings et al., 2007].

Sericea lespedeza has undoubtedly become a difficult-to-suppress, unwanted plant in some situations. It has also been a difficult-to-obtain, highly desired plant in other situations. Some currently undesired populations of this plant were deliberately planted. Others, as with previously mentioned Conservation Reserve Program plantings in Kansas [Kilgore et al., undated], were planted but inadvertently. As also previously mentioned, sericea lespedeza has the ability to volunteer or grow where not previously planted. Even with this potential, Munger [2004] noted that it most often occurs on or near sites where it was planted previously. Munger [2004] further noted that there is lack of a documented long-distance seed dispersal mechanism for this legume with inadequate information about seed dispersal not related to anthropogenic activities. Along with potential spread through hay contaminated with the seed, it has been stated that seed is spread by animals [Vermeire et al., 1998]. Ohlenbusch et al. [2001] noted that quail consume the seed and deer browse the plant in winter when seed can be present. Cattle consume the plant primarily early in the growing season long before plant reproductive development and presence of seed. Considerable progress in understanding aspects of potential invasiveness could be made with some rather simple evaluations of feces samples from various animals with access to sericea lespedeza populations following seed maturation. If substantial passage of viable seed through the digestive tract of domestic livestock readily occurs, grazing of pastures during times of seed availablilty could be managed appropriately. Certainly under such conditions, cattle should not be moved at this time from a pasture with this plant to pastures without it (if spread of the plant is not desired).

Due to widespread planting over several decades and general lack of documentation of such plantings, it is difficult to unequivocally determine that discovery of sericea lespedeza at a particular site is due to non-anthropogenic dispersal. Even recent familiarity with the plant community does not preclude long-term presence as a component of a soil seedbank. Nonetheless, controlled experiments have been used to evaluate population dynamics of sericea lespedeza and propose that sericea lespedeza is invading rangelands of the southern

Great Plains at a rate of almost 2% per year [Cummings et al., 2007]. Treatments including herbicides or burning and grazing reduced only the rate of spread of sericea lespedeza. In contrast, small plots of sericea lespedeza in Louisiana allowed to grow undisturbed and maintain a dense canopy during a 7-year period recruited new plants only within the plot area and around the immediate edges of original plants despite substantial seed production [Pitman, 2006]. No evidence of spread or potential dispersal was detected despite presence of adjacent recently disturbed sites. Such a contrast suggests that grazing livestock, which were not present on the Louisiana site, could be a key aspect of the increase on Oklahoma rangelands. Alternatively, initial dispersal by anthropogenic means may have produced a widespread but sparse population which is now increasing from seed produced on site.

Both the potential seed dispersal mechanisms and the ecosystems vulnerable to invasion need further clarification. Since many current forests in the Southeast developed on retired farmlands, soil seedbanks may include sericea lespedeza seed from previous plantings which were shaded out as forest canopies closed. Greater knowledge of soil seedbanks is needed to determine if such long-term seed survival of this species occurs. Perhaps most of the unwanted sericea lespedeza now present in the Southeast was actually planted. Whether this is the case or not, tremendous quantities of seed are now available in some locations for further distribution by either unintended human activity or natural biotic or abiotic dispersal. Undoubtedly some seed on the soil surface is moved by runoff following high-rainfall events. Some transfer of seed likely occurs within vulnerable ecosystems in association with livestock and wildlife activity. If these are only secondary or minor sources of seed dispersal to new areas and human activity is the major factor, approaches to control the spread can be focused on human activities. Populations resulting from planting seed on prepared seedbeds on Appalachian minespoil or Great Plains rangelands hardly substantiate invasiveness of even an unwanted and difficult-to-remove dominant species. Approaches appropriate and resources justified for public investment in control of sericea lespedeza may need further assessment in at least some instances. Some populations which appear to present problems may be only temporary in forest ecosystems where closure of overstory canopies can suppress and perhaps even eliminate sericea lespedeza populations. Differences in invasiveness and persistence among varieties, which differ substantially in such characteristics as tannin concentration, leafiness, nutritive value, and palatability, have apparently not been assessed. Effective strategies for sustainable control of existing sericea lespedeza populations and approaches to restrict additional unwanted spread appear to require further research for adequate efficacy of recommendations in many instances.

Mimosa

In contrast to the situation with kudzu and sericea lespedeza, which were widely planted for forage in the Southeast before concerns developed about their invasiveness, the invasiveness of mimosa is currently recognized with only potential as a forage plant established. An apparently obvious conclusion is that this potential usefulness is offset by the risk, which should therefore preclude any further consideration for forage use. A basis for continuing recognition of the forage potential of mimosa, however, is provided by the promise of future technological developments. Ranney [2004] presented the concept of research in progress to develop seedless triploid varieties of several horticultural species including mimosa. Such varieties can be produced by hybridizing a tetraploid with a diploid of the species to produce seed for propagation of a non-seed-producing triploid generation. Triploids can also be produced by generation of plants from naturally triploid endosperm tissue. In either case, further multiplication for field plantings of mimosa can be made vegetatively by rooting stem cuttings for generation of transplants. Thus, depending on source of future seedless mimosa varieties, seed for a terminal generation or vegetative propagation of transplants could provide pastures of non-invasive mimosa.

A seed propagated variety of non-invasive mimosa would likely be more economical than one dependent on vegetative propagation. A sustainable system would, however, require more intensive management than typical for perennial grasses in the region to maintain the necessary balance between excessive woody growth and defoliation-induced mortality [Pitman, 2008]. The comparatively high nutritive value of mimosa and potential for use as grazed pasture or harvested forage in systems similar to tropical cut-and-carry methods may provide opportunity for novel, intensively managed, economically viable livestock agriculture with limited external inputs. Such possibilities are sufficiently promising to be the basis for a U.S. patent on one such pasture-based system titled "Process for Propagation and Utilization of Mimosa" [Bransby et al., 2003]. Alley-cropping and other agroforestry systems based on mimosa have also been evaluated [Matta-Machado and Jordan, 1995; Rhoades et al., 1998; Addlestone et al., 1999; Burner et al., 2008].

The various pests of mimosa in the Southeast may, therefore, be considered beneficial for control of existing unwanted populations or potential problems for a rather intensively managed mimosa crop. Nematodes and Fusarium wilt [McArdle and Santamour, 1986], mimosa webworm [Santamour, 1990], the beetle *Agrilus subrobustus* [Westcott, 2007], and the psyllid *Acizzia jamatonica* [Ulyshen and Miller, 2007] are among an apparently growing list of mimosa pests. Abundant seed production, wide adaptation, root sprouting of damaged trees, and dispersal of seedpods by wind for short distances and water for greater distances contribute to the spread of this species in disturbed areas and along forest edges in the Southeast. Recommendations for control primarily involve herbicide applications, including injection of individual trees when practical [Miller, 2003]. Competitive ability under a forest canopy is poor [Smith et al., 2001], and grazing prevents establishment of seedlings in pastures. Even though grazing typically prevents establishment of mimosa, grazing does not provide a realistic short-term remedy for established trees. Smith et al. [2001] noted that mimosa does not often persist for more than 30 years without renewed disturbance. In Louisiana, mimosa populations dominating small areas are often in obvious decline within 10 years. Plants defoliated periodically, such as with infrequent roadside mowing, typically regrow vigorously and do not exhibit the symptoms of decline apparent on plants allowed to grow to maturity. This regrowth vigor is a key aspect of the forage potential of the species. Maintaining such a vegetative growth stage restricts both pest-induced mortality and seed production. Thus, mimosa under intensive management as a grazed forage crop would have limited opportunity to contribute to invasiveness. As often occurs with perennial plants however, cessation of management could provide a source of seed for new invasions.

Tropical Species

Large numbers of tropical legume species were evaluated for adaptation and potential usefulness as forage plants in Florida, primarily in the 1970s and 1980s. One area of focus for these evaluations was the Spodosol (flatwoods) region, which has been a major cattle production area. Additional evaluations were conducted on the central ridge where citrus crops predominated. Early evaluations were also conducted further north where winter temperatures restricted most of the germplasm evaluated. Soils throughout these areas where pasture legumes were especially needed were infertile, acid, and sandy. At least a few mild winter frosts each year characterize the entire region and greatly limit adaptation of tropical legumes. Seasonally waterlogged soils distinguish the Spodosol portion and further limit the range of species adapted.

On the well-drained, non-waterlogged soils, rhizoma peanut was identified in early evaluations as adapted and was developed largely as a high-value hay crop. Although a tropical plant, cold tolerance is sufficient for this legume to be useful into Alabama and Georgia as well as in northern Florida and west at least into Louisiana. The species is vegetatively propagated for commercial plantings, and the small amount of seed produced is below the soil surface. Thus, this plant does not present an invasion hazard. It has been commercially successful but only on a small scale. On the Spodosols, the genus *Aeschynomene* has been widely planted for forage with the annual *A. americana* predominating. Although some species of the genus have potential for weediness, plantings in Florida have more often been characterized by establishment difficulties and lack of persistence. Limited commercial use of the short-lived perennial phasey bean (*Macroptilium lathyroides*) has been made with seed harvested from naturalized populations on an opportunistic basis. Plantings of this species did not provide sustained pasture stands. The only relatively sustainable pasture species for the Spodosols has been carpon desmodium (*Desmodium heterocarpon*), which has often been difficult to establish rather than invasive. Leucaena (*Leucaena leucocephala*) showed promise particularly on well drained soils where both low soil pH and winter frosts were sufficient limitations to prevent commercial adoption. Seasonally wet soils, low pH, and excessive defoliation of seedlings by deer were limiting constraints despite considerable research effort with this species on the Spodosols.

In contrast to results from forage development efforts, leucaena is considered invasive on limestone outcroppings in extreme southern Florida. These sites are naturally hardwood hammock forests where leucaena invades the forest edges [Horvitz et al., 1998]. It has been proposed that this species is among a group of hardseeded, non-native leguminous trees which dominate soil seedbanks and displace native species with similar functional roles through competition among seedlings following disturbance [Horvitz et al., 1998]. Leucaena has been marketed in southern Florida for decades as a landscape plant under the name lead tree and is now on the voluntary "do not sell" list developed by the Florida Nursery Growers and Landscape Association and the Florida Exotic Pest Plant Council [Caton, 2005]. Control is difficult in natural areas where invasion problems are recognized, and available herbicide treatments have not been consistently effective [Langeland and Stocker, 2001].

Phasey bean, mentioned above as non-persistent as a forage plant, can be weedy in some situations and is sometimes considered invasive. It is an early colonizer with considerable seed production potential. In addition to functioning as an early successional plant, phasey bean is highly palatable to grazing livestock. Deer selectively browse immature pods, which limits seed production. In addition to rather sporadic occurrence in localized colonies on Spodosol soils following disturbance, phasey bean can be particularly weedy in disturbed areas of the southern Florida landscape mosaic of agricultural land, residential areas, pine savannas, and hardwood hammock forests. It is not a recognized, persistent threat to natural areas, but it can be a highly visible undesired temporary component of disturbed sites.

Conclusion

Research programs to improve forage production for the southern livestock industry have been in progress across the region for more than a century. The characteristic infertile soils and associated low quality, summer-growing grasses indicate potential for substantial benefit from the perennial warm-season legume plant functional group. Considering the tremendous number of warm-season or summer-growing legumes described in the taxonomic literature, rate of success in development of existing species for use as commercially propagated forage plants is extremely low. This plant group includes a tremendous range in growth forms from trees and vines to herbs. An extensive range in primary propagation mechanisms from primarily climbing viney spread to obligate seed multiplication is represented. These wide ranges are included among the few species of value or potential value for forage and among those considered invasive. The potential for invasiveness is closely associated with success as a forage plant due to similar requirements for high levels of success in the two roles. Conversion of almost the entire southern landscape to agricultural uses early in the last century left little recognizable need to consider potential escape of crop plants. Kudzu presents a rather drastic illustration of unintended and unexpected consequences, with no real solution readily apparent. Control efforts must be made, but these must be more strategically based than many past unsuccessful efforts. In many situations, use of grazing animals appears to provide a cost-effective component for a control strategy. Control of kudzu in commercial forests, natural areas, and urban landscapes must be addressed.

The additional perennial warm-season legumes potentially useful as forage plants must be considered in light of the unexpected results from kudzu. Sericea lespedeza can become a problem within the landscapes where it is now used as a forage and conservation plant. Effective strategies for dealing with the existing problem of unwanted stands and preventing future similar situations are needed. The limited available resources for control of plant pest problems in either commercial or non-commercial situations demand that priorities be set and only the highest priority plant invasion problems fully addressed. As with kudzu, some level of reduced-cost limitation of continuing spread of both sericea lespedeza and mimosa in localized situations should be attainable with strategic grazing perhaps selecting species of grazing animal according to the plant community involved. Cattle as generalist grazers may be less desirable where grasses are to be provided competitive advantage. Goats typically browse woody species in preference to grasses, while sheep can be used to some extent to target forb species over grasses and woody plants. Current information indicates that planning, timing of interventions, monitoring, and planned follow-up over periods of several years will be required to control existing undesired populations of the perennial legumes kudzu, sericea lespedeza, and mimosa. In some situations, ecological processes over extended time may be key components of economical control strategies. Also, more emphasis on developing the potential forage resources provided by native species of this plant functional group, rather than primary emphasis on introduction of novel species, appears appropriate.

References

Addlestone, B. J., Mueller, J. P., & Luginbuhl, J. M. (1999) The establishment and early growth of three leguminous tree species for use in silvopastoral systems of the southeastern USA. *Agroforestry Systems*, *44*, 253-265.

Altom, J. V., Stritzke, J. F., & Weeks, D. L. (1992) Sericea lespedeza (*Lespedeza cuneata*) control with selected postemergence herbicides. *Weed Technology*, *6*, 573-576.

Baldridge, J. D. (1957) The lespedezas: II. culture and utilization. *Advances in Agronomy*, *9*, 122-141.

Ball, D. M., Hoveland, C. S., & Lacefield, G. D. (2002) *Southern Forages* (3rd edition). Norcross, GA: Potash and Phosphate Institute.

Ball, D. M. & Mosjidis, J. A. (1995) An objective look at sericea lespedeza. *Southern Pasture and Forage Crop Improvement Conference Proceedings*, *51*, 50-55.

Bonsi, C., Rhoden, E., Woldeghebriel, A., Mount, P., Solaiman, S., Noble, R., & Paris, G. (1992) Kudzu-goat interactions-a pilot study. In: S. G. Solaiman & W. A. Hill (Eds.) *Using Goats to Manage Forest Vegetation: a Regional Inquiry* (pp. 84-88). Tuskegee, AL: Tuskegee University Agricultural Experiment Station.

Boyette, C. D., Abbas, H. K., & Walker, H. L. (2001) *Control of Kudzu with a Fungal Pathogen Derived from Myrothecium verrucaria (United States Patent 6274534).* [online] Available from: http://www.freepatentsonline.com/6274534.html

Boyette, C. D., Walker, H. L., & Abbas, H. K. (2002) Biological control of kudzu (*Pueraria lobata)* with an isolate of *Myrothecium verrucaria. Biocontrol Science and Technology*, *12*, 75-82.

Brandon, A. L., Gibson, D. J., & Middleton, B. A. (2004) Mechanisms for dominance in an early successional old field by the invasive non-native *Lespedeza cuneata* (Dum. Cours.) G. Don. *Biological Invasions*, *6*, 483-493.

Bransby, D. I. (1993) Preliminary evaluation of *Albizia jusibrissin* as a woody warm-season forage legume with high cold tolerance. *Proceedings of International Grassland Congress, XVII*, 2057-2058.

Bransby, D. I., Duke, S. R., & Krishnagopalan, G. A. (2003) *Process for Propagation and Utilization of Mimosa (United States Patent 20030204988)*. [online] Available from: http://www.freepatentsonline.com/20030204988.html

Britton, K. O., Orr, D., & Sun, J. (2002) Kudzu. In: *Biological Control of Invasive Plants in the Eastern United States,* USDA Forest Service Publication FHTET-2002-04. [online] Available from: http://www.invasive.org/eastern/biocontrol/25Kudza.html

Burner, D. M., Carrier, D. J., Belesky, D. P., Pote, D. H., Ares, A., & Clausen, E. C. (2008) Yield components and nutritive value of *Robinia pseudoacacia* and *Albizia julibrissin* in Arkansas, USA. *Agroforestry Systems*, *72*, 51-62.

Caton, B. P. (2005) *Availability in Florida Nurseries of Invasive Plants on a Voluntary "Do Not Sell" List.* Raleigh, NC: USDA-APHIS-Plant Protection and Quarantine.

Cope, W. A. (1957) The lespedezas: III. lespedeza breeding and improvement. *Advances in Agronomy*, *9*, 142-157.

Cummings, D. C., Fuhlendorf, S. D., & Engle, D. M. (2007) Is altering grazing selectivity of invasive forage species with patch burning more effective than herbicide treatments? *Rangeland Ecology and Management*, *60*, 253-260.

Demers, C. & Long, A. (2002) *Controlling Invasive Exotic Plants in North Florida Forests.* Gainesville, FL: University of Florida. [online] Available from: http://edis.ifas.ufl.edu/pdffiles/FR/FR13300.pdf

Dove, D., Wolf, D., & Zipper, C. (1997) *Conversion of sericea lespedeza-dominated vegetation to quality forages for livestock use.* Blacksburg, VA: Virginia Cooperative Extension.

Duggar, J. F. (1925) *Southern Forage Crops*. New York: The Macmillan Company.

Everest, J. W., Miller, J. H., Ball, D. M., & Patterson, M. (1999) *Kudzu in Alabama: History, Uses, and Control*. Auburn, AL: Alabama Cooperative Extension System.

Forseth, Jr., I. N. & Innis, A. F. (2004) Kudzu (*Pueraria montana*): history, physiology, and ecology combine to make a major ecosystem threat. *Critical Reviews in Plant Sciences*, *23*, 401-413.

Frye, M. J., Hough-Goldstein, J., & Sun, J. (2007) Biology and preliminary host range assessment of two potential kudzu biological control agents. *Environmental Entomology*, *36*, 1430-1440.

Harrington, T. B., Rader-Dixon, L. T., & Taylor, Jr., J. W. (2003) Kudzu (*Pueraria montana*) community responses to herbicides, burning, and high-density loblolly pine. *Weed Science*, *51*, 965-974.

Henson, P. R. (1957) The lespedezas: I. the origin, history, and development of lespedeza in the United States. *Advances in Agronomy*, *9*, 113-122.

Hipps, C. B. (1994) Kudzu, a vegetable menace that started out as a good idea. *Horticulture*, *72*, 36-39.

Horvitz, C. C., Pascarella, J. B., McMann, S., Freedman, A., & Hofstetter, R. H. (1998) Functional roles of invasive non-indigenous plants in hurricane-affected subtropical hardwood forests. *Ecological Applications*, *8*, 947-974.

Jonsson, P. (2003) In curbing kudzu, goats may be man's best friend. *The Christian Science Monitor*, [online] Available from: http://www.csmonitor.com/2003/1021/p01s02-ussc.htm

Jordan, M. J., Lund, B., & Jacobs, W. A. (2002) Effects of mowing, herbicide and fire on *Artemisia vulgaris*, *Lespedeza cuneata* and *Euphorbia cyparissias* at the Hempstead Plains grassland, Long Island, New York. Albany, NY: New York State Museum.

Kilgore, G., Davidson, J., & Fick, W. H. (undated) *Sericea Lespedeza History* (Forage Facts Publication Series). Manhattan, KS: Kansas State University.

Koger, C. H. & Stritzke, J. (2004) Sericea lespedeza (*Lespedeza cuneata*). *Southern Weed Science Society Proceedings*, [online] Available from: http://www.ars.usda.gov/research/publications/publications.htm?seq_no_115=145425&pf=1

Koger, C. H., Stritzke, J. F., & Cummings, D. C. (2002) Control of sericea lespedeza (*Lespedeza cuneata*) with triclopyr, fluroxypyr, and metsulfuron. *Weed Technology*, *16*, 893-900.

Landers, J. L. & Mueller, B. S. (1986) *Bobwhite Quail Management: a Habitat Approach.* Tallahassee, FL: Tall Timbers Research Station.

Langeland, K. A. & Stocker, R. K. (2001) *Control of Non-native Plants in Natural Areas of Florida*. Gainesville, FL: University of Florida, IFAS Extension.

Li, J., Bruce, L. M., Byrd, J. & Barnett, J. (2001) Automated detection of Pueraria montana (kudzu) through Haar analysis of hyperspectral reflectance data. In: *IEEE International Geoscience and Remote Sensing Symposium* (pp. 2247-2249). Piscataway, NJ: Institute of Electrical and Electronic Engineers.

Luginbuhl, J., Green, J. T., Poore, M. H., & Mueller, J. P. (1996) *Use of Goats as Biological Agents for the Control of Unwanted Vegetation*. Raleigh, NC: North Carolina State University. [online] Available from: http://www.cals.ncsu.edu/an_sci/extension/animal/meatgoat/MGVeget.htm

Matlack, G. R. (2002) Exotic plant species in Mississippi, USA: critical issues in management and research. *Natural Areas Journal*, *22*, 241-247.

Matta-Machado, R. P. & Jordan, C. F. (1995) Nutrient dynamics during the first three years of an alleycropping agroecosystem in southern USA. *Agroforestry Systems*, *30*, 351-362.

McArdle, A. J. & Santamour, F. S. (1986) Screening mimosa (*Albizia julibrissin*) seedlings for resistance to nematodes and Fusarium wilt. *Plant Disease*, *70*, 249-251.

Miller, J. H. (1996) Kudzu eradication and management. In: D. Hoots & J. Baldwin (Eds.), *Kudzu: the Vine to Love or Hate* (pp. 137-149). Kodak, TN: Suntop Press.

Miller, J. H. (2003) *Nonnative Invasive Plants of Southern Forests: a Field Guide for Identification and Control*. Asheville, NC: Southern Research Station, USDA Forest Service.

Moorhead, D. J. & Johnson, K. D. (1998) *Controlling Kudzu in CRP Stands*. Athens, GA: Georgia Cooperative Extension Service.

Munger, G. T. (2004) Lespedeza cuneata. In: *Fire Effects Information System*. USDA Forest Service, [online] Available from: http://www.fs.fed.us/database/feis/plants/forb/lescun/all.html

Ohlenbusch, P. D. & T. Bidwell. (2001) *Sericea Lespedeza: History, Characteristics and Identification*. Missouri Department of Conservation [online] Available from: http://mdc.mo.gov/landown/grass/sericea/

Pappert, R. A., Hamrick, J. L., & Donovan, L. A. (2000) Genetic variation in *Pueraria lobata* (Fabaceae), an introduced, clonal, invasive plant of the southeastern United States. *American Journal of Botany*, *87*, 1240-1245.

Pitman, W. D. (2006) Stand characteristics of sericea lespedeza on the Louisiana Coastal Plain. *Agriculture, Ecosystems and Environment*, *115*, 295-298.

Pitman, W. D. (2008). Establishment and regrowth responses of *Albizia julibrissin* on Louisiana USA Coastal Plain soils. *Agroforestry Systems* 74:259-266.

Ranney, T. G. (2004) Population control: developing non-invasive nursery crops. *International Plant Propagator's Society Combined Proceedings*, *54*, 604-607.

Rhoades, C. C., Nissen, T. M., & Kettler, J. S. (1998) Soil nitrogen dynamics in alley cropping and no-till systems on Ultisols of the Georgia Piedmont, USA. *Agroforestry Systems*, *39*, 31-44.

Santamour, Jr., F. S. (1990) Trees for urban planting: diversity, uniformity, and common sense. *Metropolitan Tree Improvement Alliance (METRIA) Conference Proceedings*, *7*, 57-65.

Sasek, T. W. & Strain, B. R. (1988) Effects of carbon dioxide enrichment on the growth and morphology of kudzu (*Pueraria lobata*). *Weed Science*, *36*, 28-36.

Sasek, T. W. & Strain, B. R. (1989) Effects of carbon dioxide enrichment on the expansion and size of kudzu (*Pueraria lobata*) leaves. *Weed Science*, *37*, 23-28.

Sasek, T. W. & Strain, B. R. (1990) Implications of atmospheric CO_2 enrichment and climatic change for the geographical distribution of 2 introduced vines in the USA. *Climate Change*, *16*, 31-51.

Smith, D., Wofford, B. E., & Clebsch, E. E. C. (2001) *Vegetation Management Evaluation and Exotics Survey, Part 1, Invasive Exotic Plant Species on Tapoco Project Lands*. Nashville, TN: The Nature Conservancy of Tennessee.

Sturkie, D. G. (1951) Hay and pasture seedings for the humid South. In: H. D. Hughes, M. E. Heath, & D. S. Metcalfe (Eds.), *Forages: the Science of Grassland Agriculture* (pp. 488-497). Ames, IA: Iowa State College Press.

Sun, J. H., Li, Z., Jewett, D. K., Britton, K. O., Ye, W. H., & Ge, X. (2005) Genetic diversity of *Pueraria lobata* (kudzu) and closely related taxa as revealed by inter-simple sequence repeat analysis. *Weed Research*, *45*, 255-260.

Ulyshen, M. D. & Miller, D. R. (2007) First record of *Acizzia jamatonica* (Hemiptera: Psyllidae) in North America: friend or foe? *Florida Entomologist*, *90*, 573.

Vermeire, L. T., Bidwell, T. G., & Stritzke, J. (1998) *Ecology and Management of Sericea Lespedeza*, (OSU Extension Facts F-2874). Stillwater, OK: Oklahoma Cooperative Extension Service.

Webster, C. R., Jenkins, M. A., & Jose, S. (2006) Woody invaders and the challenges they pose to forest ecosystems in the eastern United States. *Journal of Forestry*, *104*, 366-374.

Westcott, R. L. (2007) The exotic *Agrilus subrobustus* (Coleoptera: Buprestidae) is found in northern Georgia. *The Coleopterists Bulletin*, *61*, 111-112.

In: Invasive Species: Detection, Impact and Control
Editors: C.P. Wilcox and R.B. Turpin
ISBN 978-1-60692-252-1

Chapter 8

INVASIVE SWIMMING CRABS: DEVELOPMENT OF ERADICATING TRAPPING GEAR AND METHODS

Miguel Vazquez Archdale*

Field of Fisheries Engineering, Faculty of Fisheries, Kagoshima University,
Shimoarata 4-50-20, Kagoshima City 890-0056, Japan

ABSTRACT

Making simple improvements on traditional fishing gear and methods can be applied to develop effective eradicating tools that can be integrated with other strategies used to combat invasive aquatic species. Swimming crabs are the target choice; they travel widely by swimming, attached to the hulls of ships or as larvae suspended in ballast water. They represent a valuable fisheries resource and are consumed as food. They are however also carriers of pathogens that may damage fisheries and aquaculture industries, which results in the need of effective eradication methods.

Traps make good eradicating gear because they are light, inexpensive, small, easily stacked onboard, keep the catch alive, and permit the release of non-target organisms undamaged. They are passive gear, so the target organism has to be lured into the traps with the promise of food, shelter, company or even a mating partner.

Improving luring methods, such as common baits or pheromone emitting decoys, is just as important as research on new trap design for increasing the catch. Determining the taste preferences of the target crab, and making bait combinations of fish plus sugarcane have demonstrated a synergistic effect that doubled the catch. Alternative bait made from fish mince inside teabags was as effective as ordinary bait, and allowed enrichment of the bait with attractants and the use of fish byproducts as raw material, which reduces cost and conserves valuable food resources. Pheromones offer attracting potential, have been found in swimming crabs and are emitted by the live decoys employed in their trap fishery. This fishing method eliminates the non-target catch by attracting only crab conspecifics. Specially designed traps containing live decoy crabs were more effective than baited traps, and eliminated unwanted organisms that feed on bait.

When designing eradicating gear, the quantity of the catch is more important than its size or quality; consequently, traps must be fitted with smaller meshed netting to

* E-mail: miguel@fish.kagoshima-u.ac.jp

additionally retain juvenile crabs. Observations on crab behavior around different trap designs showed that they search larger areas around traps with oval bases than those with rectangular ones, and this gives them more access to the trap entrances. Open funnel entrances allowed entry of most crabs contacting the traps and were superior to tight slit entrances because being open they permitted escape, thus reducing the non-target catch and the negative impact that lost gear causes on the aquatic resources by ghost fishing.

INTRODUCTION

Increases in ocean traffic, construction of manmade canals, and release of live animals and plants into new environments are responsible for the introduction of numerous marine organisms around the world. Among them, crustaceans account for a large proportion of this total amount, and have been transported attached to the hulls of ships and in ballast water, intentionally released into new environments to enrich fisheries resources, or made their own way by swimming across new canals connecting water bodies that were geographically isolated in the past. These species, finding a new habitat devoid of ecological checks, such as predators, diseases or parasites, occasionally proliferate and become a nuisance by competing with local organisms for space and resources. They can potentially damage important fisheries and aquaculture industries by targeting local species as food prey or by transmitting new pathogens to which the local species have no natural resistance.

Humans have dealt with invading organisms in the past by adopting several strategies to contain their spread, but all attempts to eradicate them must begin with a detailed study of their ecology and life cycle, in order to determine the weakest points where pressure can be applied (Burdick 2005). Simple approaches have been to introduce a specific predator or parasite of the invading species hoping to limit its growth or hinder its reproduction, to attempt to remove or kill it using traps or poisons, or by encouraging hunters and fishers to catch them for a small reward. The latter may offer possibilities with some decapod crustaceans because they are already fished and represent a valuable fisheries resource. Humans consume crabs for food; consequently, sometimes developing a new fishery can be applied to reduce their numbers and enrich the local diet, particularly if there is one already in their country of origin. In some cases the invading crab species may have been studied in its place of origin; in that case, searching for literature and interviewing fishers or researchers may highlight some possible solutions.

The target swimming crabs used as models in this chapter, the Japanese shore swimming crab *Charybdis japonica* and the blue swimming crab *Portunus pelagicus,* are commercially exploited in Japanese waters, which are part of their natural habitat (Fig. 1). These crabs are also considered invasive, *C. japonica* in the coastal waters of Australia and New Zealand, and *P. pelagicus* in the eastern Mediterranean Sea. For this reason, studying commercial fishing gear that is currently being employed to capture them can help determine the most efficient trap type, and improvements in its design can serve to create a new eradication trap. Furthermore, because of the possibility of losing traps in the area where they are deployed, ways to minimize the negative impact they could have if lost in the fishing ground should be incorporated into the new design.

Figure 1. Japanese shore swimming crab *Charybdis japonica* (top) and blue swimming crab *Portunus pelagicus* (bottom).

Reasons for Choosing Traps as Eradicating Tools

Humans are responsible for the extinction of numerous animals from prehistoric times, and hunting communities have developed extremely effective tools and methods to capture their prey. Fishers represent one of the few remaining legacies of this hunting past, and their technological progress and mentality are blamed for the unsustainable management of many aquatic resources that has resulted in the overexploitation of several commercially important fisheries. Applying this mentality and technology, which has proved so effective, may offer positive aspects if used to eradicate invasive species or at least to contain their spread.

Of all the fishing methods fishers employ, traps have a long history of tested effectiveness as harvesting gear for aquatic organisms, and there are extensive examples of commercial fisheries still employing them as their preferred gear. Traps are regularly used to capture crustaceans because they are cheap, small and light, easy to operate, occupy little deck space in the fishing boat, are selective towards the catch, and will maintain a large proportion of the captured organisms alive and permit the return of undersized and unwanted species unharmed back into the water. Traps are a passive fishing gear, where the target organism has to be lured into them by using bait. This relies on the locomotive power of the target animals, which have to approach the gear by themselves; therefore, this technique requires less energy and fuel than active fishing methods used to catch crustaceans, such as trawling. Furthermore, small fishing gear has lower operational costs, can be hauled manually or with smaller line haulers than other gear, and sometimes can be slightly modified so existing gear can be adapted for eradication purposes.

The Japanese collapsible crab traps introduced in this chapter are very effective fishing gear and some designs are already in use for eradication, sampling and research (Behrens Yamada et al. 2006); consequently, improvement in their design can be applied to develop better traps that can be integrated with other eradication strategies (Lafferty and

Kuris 1996); like the creation of a new fishery where there is no tradition of eating these edible crabs; improving habitat, spawning substrate and numbers of natural predators; and perhaps the use of species-specific parasites which can cause infertility or reduce growth rates of the introduced crabs. Collapsible traps occupy less space on deck because they are two-dimensional and can be laid flat until they are assembled, just before being tossed into the sea (Fig. 2). For this reason larger numbers can be carried onboard than those of conventional rigid trap design. Even sport fishers can easily adapt their boats to this type of fishery.

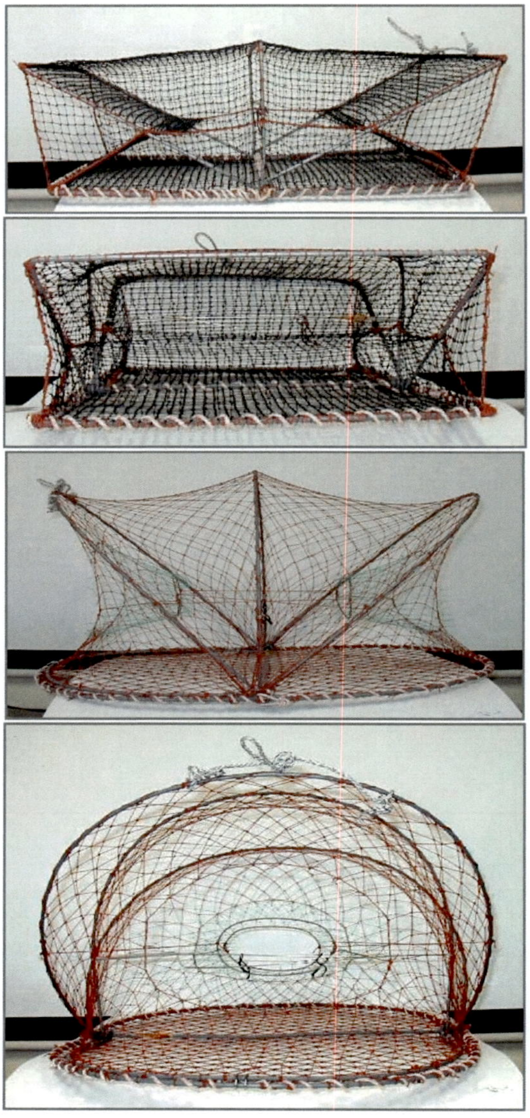

Figure 2. Side views of box trap with slit entrances (top 2) and dome trap with open funnels (bottom 2).

Trap Operation

Traps are usually baited with mackerel or sardine when the fishing boat is in the port just before going out to sea. Fishers rely on bait and the acute sense of smell of crustaceans to lure them towards the traps. The rigging used during crab fishing usually consists of individual traps fastened to a branch line, which connects them to a bottom longline. Traps are spaced at regular intervals (15 m for swimming crab traps), to prevent tangling and to eliminate competition between bait. The bottom longline is tied at both extremes to an anchor or weight, and to another line connected to a marker buoy, which serves to locate and haul the gear.

Capture Process

Traps employed to catch crustaceans are usually set before sunset and collected after dawn, so that the time they spend in the fishing ground (soaking time) coincides with the active phase of the target species. Swimming crabs are mostly nocturnal, though they also show short periods of activity during the day, and as a result have to rely mostly on their senses of smell and taste to find food or prey.

The capture process of an aquatic animal approaching a trap has been summarized as consisting of several behavioral stages (Furevik 1994), which start with distant detection, and continue with approach, contact, near ingress behavior, sometimes entry, and end with capture and/or escape. The first two, detection and approach are mediated by the sense of smell of the organism and are influenced by the distribution of the odor emanating from the trap as the current carries it downstream. Fishers can place bait or live crab decoys in their traps to attract other crabs towards them. I will discuss the lures they employ in the following sections.

Luring the Crab to the Trap

Bait

Many fishers use fish of no commercial value or "trash fish" as bait, but animal by-products such as viscera, skin or heads, salted or dried fish, smoked cormorant, frogs, chicken offal and many other things can be used as well. Commercial trap fishers usually prefer to use frozen fish bait, like mackerel or sardine, because from past experience they know that to have a good catch they need high quality bait. Some studies have examined the active ingredients found in bait that are responsible for the attraction that induces animals to move into the trap. Swimming crabs are scavengers that rely heavily on their sense of smell for locating their food. Their nose-equivalent is a small appendage resembling an antenna called the "antennule", and the two they possess are located just between their eyes (Fig. 3). Their entire body surface is also lined with chemical and mechanical receptors, so that walking legs, mouthparts and claws act as sensors that help them locate and handle their prey. Water-soluble compounds from damaged tissues, excretions and decaying organisms act as strong stimuli and affect crab feeding behavior, and this explains why fishers employ bait to attract crabs into traps.

Research conducted to identify individual attracting substances found in decapod crustacean food and prey has been carried out in the past, mostly by bioassay and electrophysiological experiments. Isolating stimulating compounds from natural prey organisms and testing them with live decapods, or appendage preparations where the receptors are still viable, have elucidated that substances of low molecular weight, such as amino acids, saccharides (sugars), nucleotides and ammonium compounds, are mostly responsible for the food searching responses elicited by the bait (Ache 1982, Carr 1988). Other findings have documented that different species of decapods respond to different mixtures of these compounds, and that the effect of mixing individual substances is synergistic.

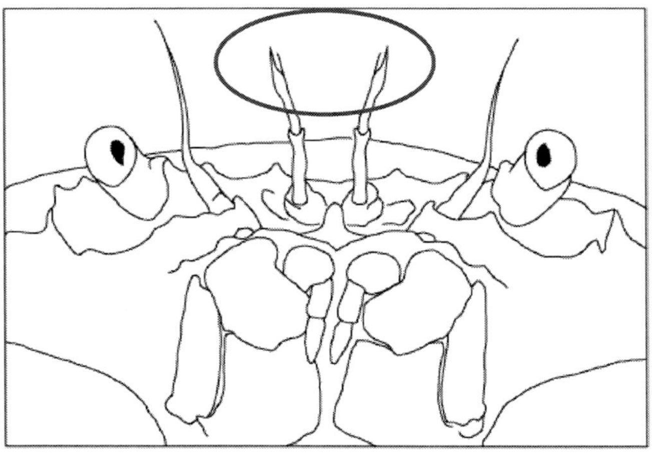

Figure 3. Frontal view of *P. pelagicus* showing mouthparts, eyes, antenna and the position of antennules (inside the circle).

By studying the behavior of swimming crabs placed in aquaria after adding a known volume of fish extract to the seawater in which they are held, a characteristic food searching sequence can be observed. Crabs exposed to fish extract commonly show an increase in the rate of antennule flicking (equivalent to sniffing through our nose), move their mouthparts, and crawl searching for food around the aquarium while prodding its sandy bottom with legs and claws trying to find the food item. Once this behavioral sequence is observed, identifying stimulating substances can be easily accomplished by adding individual chemicals diluted in seawater at increasing concentrations with a dropper and observing which elicit the same responses as the fish extract. Following this method, individual stimulating substances can be determined, together with their sensitivity thresholds. Bioassay experiments using the swimming crabs *P. pelagicus* and *Charybis feriata* have shown that the most stimulatory amino acids were alanine, glycine and serine within the 2×10^{-7} - 2×10^{-4} mol/l range; while the saccharides galactose and glucose were more stimulant than these amino acids at the same concentrations (Archdale and Nakamura, 1992). These levels are only slightly above the low ambient amino acid levels found in seawater, which are in the range of 10^{-8} mol/l (Carr 1988), and they confirm that crabs can detect minor changes in the chemical composition of the water surrounding them. These same amino acids have been identified by chemical analysis in the prey organisms of swimming crabs, which consist mainly of other crustaceans, molluscs and polychaete worms. Saccharides are present in both the blood of fish and hemolymph of crustaceans. Though similar responses showing the dual importance of sugars in crab chemoreception have been observed in the porcelain, ghost and fiddler crabs, possible applications of using sweet bait were not investigated until 1995 (Kawamura et al. 1995), when sugarcane was tried as an alternative bait for crab traps during fishing trials.

Sugarcane and Fish Bait Combinations

After determining which substances are responsible for the stimulating nature of bait, the possibility of making it more attractive should be considered. In the previous section, sugars where found to be highly attracting to swimming crabs. Sugarcane is a cheap and readily available agricultural crop (Fig. 4), which can be easily grown in tropical climates.

Figure 4. Sugarcane

The possibility of using short sections of split sugarcane as alternative bait for crab traps was tested during fishing trials in Kagoshima, southern Japan (Kawamura et al. 1995). The baiting treatments applied to the traps were fish bait, sugarcane, a combination of the two, and no bait (Fig, 5 top). Swimming crab catches of *P. pelagicus* were almost doubled and that of *C. japonica* more than tripled in traps baited with the fish-sugarcane combination, while sugarcane alone was not effective. The increase in attractiveness of the bait combination was probably due to a synergistic effect resulting from the mixture of attractive sugars and amino acids present in both fish and sugarcane. Further trials carried out in Panay Island, the Philippines, using the same treatments for trap bait also confirmed the effectiveness of the bait combination, which caught 48% more crabs than those baited with fish alone (Fig. 5 bottom) (Anraku et al. 2001). During the study in the Philippines, sugarcane baited traps still caught one-third the number of crabs caught by fish baited traps, efficient enough to justify its application, as reported by some of the fishers that follow this practice locally.

Sweet substances can enhance the effectiveness of ordinary fish baits for traps targeting swimming crabs, and might be used in a purer form to enrich existing fish bait and boost effectiveness. The next section will examine the possibility of enriching current fish bait with sugar.

Figure 5. Baiting treatments: sugarcane, fish and combination (top 3) and fishing operation using bamboo basket traps in Panay Island, the Philippines (bottom).

Fish Mince in Teabag

Enriching the bait with attractants, in the same way aquaculture feeds are supplemented with vitamins and minerals, has become a new possibility. Teabags have been used to bind fish processed into mince and may be used as a cheaper alternative to conventional fish bait (Vazquez Archdale et al. 2008). The fish was minced using a meat chopper and packed into large teabags and used as alternative bait (Fig. 6). Traps baited with fish mince teabags caught as many crabs as those using the same amount of ordinary unprocessed fish bait. The remains of the fish bait found in the traps indicated that crabs prefer to consume the viscera of the fish employed; therefore, the possibility of making cheaper bait using only fish by-products, such as head, internal organs, skin and bones, should be tested in the future. This would save a considerable amount of edible fish meat that can be destined to become human food or animal feed.

This novel binding method was also used to investigate whether enriching the fish mince with ordinary table sugar to sweeten it could increase crab catches. The results showed no significant differences to using conventional fish bait in traps for *P. pelagicus*; but *C. japonica* and other swimming crab species disliked the sweet mince and were caught at half the rates of fish baited traps, which means that the taste preference varies depending on the crab species, and indicates that more selective crab baits may be designed in the future.

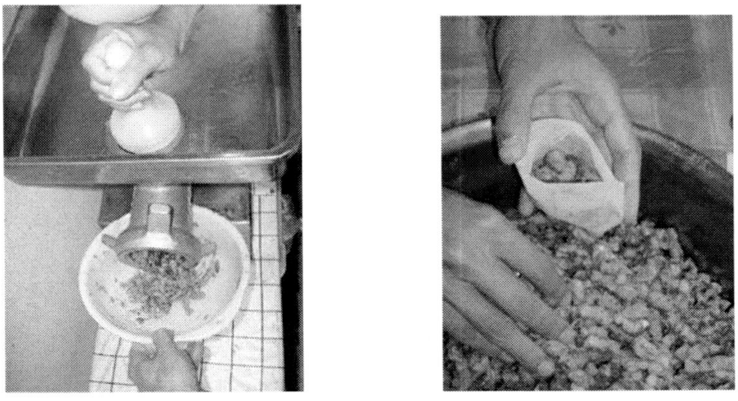

Figure 6. Mincing fish (left) and packing it into a teabag (right).

Decoys

Traditionally, two crustacean trap fisheries have used live decoys to lure conspecifics into traps. The blue crab *Callinectes sapidus* and spiny lobster *Panulirus argus* have been employed as live decoys by fishers. In this case, it is not the presence of bait that acts as an attractant, but the company of their congeners or finding a potential mating partner. The effectiveness of these decoys is still under investigation, but evidence points to the presence of sexual or aggregating pheromones in their urine or the "calls" they produce. Usually adult male blue crabs are used to attract female conspecifics into traps. This method is particularly effective during mating time, when crabs are not interested in feeding and will not enter baited traps (Van Engel 1958). Juvenile spiny lobsters, known as "shorts", are used as decoys in a similar way to attract adults. Research with lobster decoys placed in traps also confirmed their effectiveness, which was higher than the wide variety of types of bait tested (Heatwole et al. 1988).

Live decoys have several advantages over conventional fish bait, the main one being that they eliminate non-target animals from the catch, which are usually attracted into the traps by the nutritional reward found in bait. In addition, decoys can remain alive for weeks and even months while kept submerged in the water inside the traps; permitting their long-term use and at the same time reducing the waste of fisheries resources and money that bait incurs. Fish bait can only be used once and will last for a few days at most in the water due to the work of scavengers or because it rapidly decomposes.

Fishing trials were conducted to test the effectiveness of using live decoys placed in traps for the capture of *C. japonica* (Vazquez Archdale, unpublished results). At first, by simply inserting individual adult crabs into box traps with slit entrances; this entrance type acts as a one-way valve and hinders escape (Fig. 2, top). Testing the effectiveness of individual male and female crab decoys in comparison to fish bait and non-baited traps has shown that effectiveness depended on the length of time the traps remained in the water (soaking time). Trials done by soaking the traps overnight showed that fish bait was the best luring method while the decoys only attracted similar numbers to those caught in non-baited traps. On the other hand, when the traps where soaked for 4-days, the effectiveness of the female decoys proved to be six times higher than fish bait, and non-target organisms were captured at about one third the capture rate of fish bait. This suggests that there is potential to using *C. japonica* decoys or the pheromones they emit to capture conspecifics, and this could have great applications in both capture fisheries and eradication programs.

Figure 7. Decoy in trap (left) and decoy mating (right).

Several problems were encountered during the first decoy experiment; a major one was the box trap design, which permitted any octopuses in the catch to predate on the decoys. Additional fishing trials were carried out employing a new trap design with separate chambers to contain the decoys, and this was effective for eliminating predation because octopuses could not reach the crabs (Fig. 8). Furthermore, there was no decoy mortality during the month the experiment lasted, even after exposing the decoy crabs to pressure changes during trap hauling and setting to a 15-meter depth and keeping them exposed in the air for 15 minutes during the daily handling of the traps on the boat deck. Additional trials testing the effectiveness of baited and decoyed traps in a commercial crab fishing ground confirmed the effectiveness of female decoy crabs. These decoys caught an average of 0.80 crabs per decoy while only 0.25 were caught per fish bait, and this was obtained using a short 1-day soaking time. The catch of non-target organisms was also reduced five times in decoyed traps. The luring potential of female crabs was thus demonstrated and the possible application of this method for eradication purposes of invasive crabs should be further investigated.

Figure 8. Decoy in trap (top chamber) and crab catch (bottom chamber).

Recent research on another invasive crustacean, the green crab *Carcinus maenas*, is testing the possibility of manufacturing pheromone bait by binding the active components of this crab's urine into a gel matrix (Hardege et al. 2002, Behrens Yamada et al. 2006). Continuing with this line of research in the future may offer an alternative method to trapping with fish bait, which is hardly species selective. In addition, pheromone bait and decoys are only attractive to members of the species concerned, they are highly selective and greatly reduce the presence of non-target animals in the catch; therefore, their use in eradication programs may be highly valuable in the future. Important implications are that the handling stress and damage inflicted to non-target species can be eliminated, or reduced to a minimum, as a result of them not being attracted into the trap, as baited traps would.

TRAP DESIGN AND CRAB BEHAVIOR

After talking about the detection and approach stages of the capture process in trap fisheries we will proceed to examine what happens to a crab after it reaches the trap. Several studies have documented how trap design affects capture efficiency. Crabs normally crawl along the sea bottom until they encounter an odor trail emanating from a baited trap. After smelling the bait they will follow the trail upstream searching for the source. By zigzagging across this trail and sampling the odor concentrations of the substances in the water, crabs can easily locate and guide their way to the trap. After contacting the trap, this zigzag crawling will turn into short excursions to the right and left along the perimeter of the trap, and these will progress until the crab encounters an entrance that gives it access to the bait. It has also been observed that if they cannot find an entrance they will easily give up after a few attempts and look for food somewhere else.

Exploration around the Trap

By conducting observations through an underwater video camera and recording crab behavior around collapsible traps of different shapes, it was found that the area a crab explores around the perimeter of a trap varies depending on the shape of the base (Vazquez Archdale et al. 2003). An oval base (Fig. 9) gave wider average search angles (mean =173°) than a rectangular one (mean =108°), and this increases the probability of a crab finding one of the entrances. Traps targeting swimming crabs normally have two entrances located opposite to each other; therefore, search angles approaching 180° will almost guarantee that a crab finds one of the entrances. Similar results have been observed in Korea with *C. japonica*, where circular traps caught more crabs on average than rectangular ones (Kim and Ko 1987).

Behavior at Entrances

In many cases the target organism can be selected by the position and design of the trap's entrance. Collapsible traps usually have open funnel or narrow slit entrances, and these characteristics greatly affect entry rates. Observations of crab behavior underwater in different trap entrances confirmed that only 31% of the crabs contacting the slits could enter a box trap (Vazquez Archdale et al. 2003). The main reason for the low entry rate is that the spines in the crabs carapace and claws entangle with the netting material, and as a consequence they cannot enter the trap nor reach the bait; many will give up and leave the trap after a few unsuccessful attempts. On the other hand, 100% of the crabs reaching the open funnel entrances of the dome trap could enter by crawling sideways through the entrances, without encountering any obstruction.

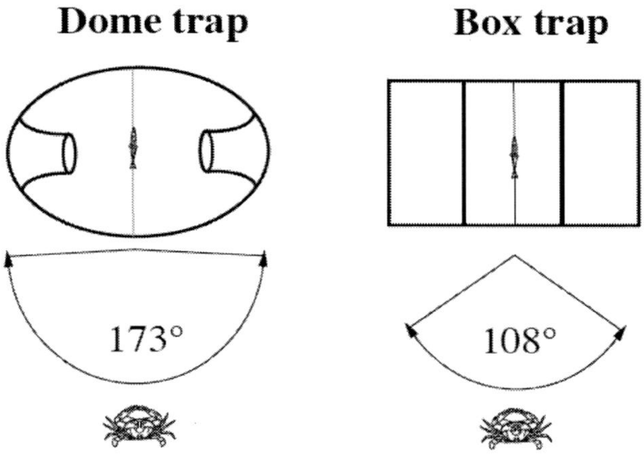

Figure 9. Crab average search angles around dome trap (left) and box trap (right) as seen from above.

Complementary observations were carried out in a large indoor tank by exposing the same 30 individual crabs to both box and dome trap designs, in order to quantify their responses and entry successes (Fig. 10). When placed in the tank only 63% of the crabs could enter the box trap in 1.5 hours, while 100% of the crabs entered the dome trap within the same time.

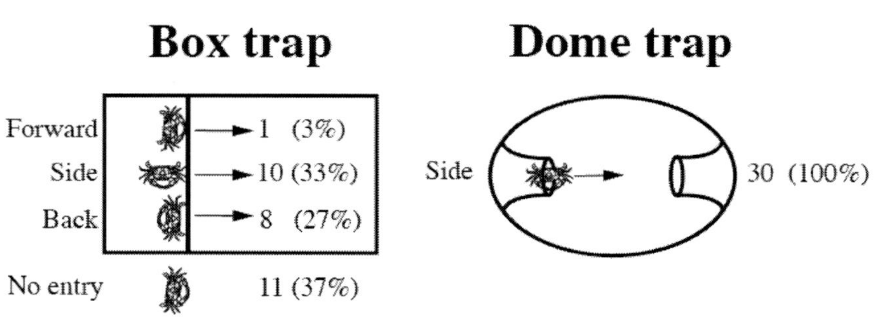

Figure 10. Types of entries observed for crabs in box trap (left) and dome trap (right).

Average times to entry after reaching the entrance were 12.3 minutes versus only 2 minutes, in box and dome traps respectively (Table 1). The average number of attempts was 13 in the box trap and only 1.3 in the dome trap, which demonstrates that crabs usually go into the latter in only one try (Vazquez Archdale et al. 2006a).

Table 1. Entry of Charybdis japonica into box and dome traps in tank experiment

Entrance type	Box trap (slit)	Dome trap (open)
No. of crabs at entrance	30	30
No. of crabs entering	19**	30
No. of attempts (avg. and range)	13 (2-45)	1.3**(1-4)
Time to entry (avg. and range)(min)	12.3 (0.31-54.90)	2.0* (0.05-28.05)

Significantly different: $*p < 0.05$ and $**p < 0.01$.

From the behavior observations it was found that the shape of the trap and design of the entrances greatly affect the ingress rates of the target crabs. Open funnel entrances fitted in dome traps were more easily encountered because the crabs searched larger areas around the perimeter of the trap. Once at the entrance, all the crabs could ingress the trap by naturally crawling sideways in the direction of the bait. Conversely, the search areas around the box traps were smaller, and in several cases the crabs could not locate an entrance during their excursions; in addition, after reaching the entrance, the netting material and narrow slits hindered their ingress efforts and resulted in many crabs giving up.

After completing the observations on crab behavior around traps of different designs and elucidating some of the factors influencing their capture process, comparative fishing trials were required to test catching performance and to determine which trap design was best for capturing crabs and would be suitable for eradication purposes.

Fishing Trials Conducted in a Pond

To test the previous behavioral results, preliminary fishing trials were conducted in a large pond using box and dome traps (Vazquez Archdale and Kuwahara 2005). Unexpectedly, the catch of the box traps was more numerous than that of the dome traps, and this was attributed to differences in mesh size, entrance type and possibly the escape rates of the captured organisms from both trap designs. Additional trials employing a newly designed dome trap fitted with smaller mesh netting (2.3 cm) as opposed to the usual mesh netting (6cm), confirmed its superiority for capturing all sizes of swimming crabs (Fig. 11) (Vazquez Archdale et al. 2006b).

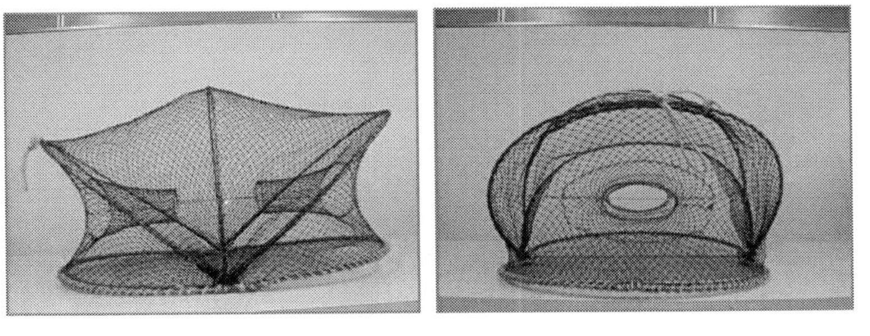

Figure 11. Side views of eradication dome trap with open funnels (mesh size 2.3 cm).

This variation of the dome design was also found ideal for eradication purposes, because the number of crabs in the catch was more numerous than the other two trap designs, regardless of the size of the crabs (Table 2).

Table 2. Catch of crabs and non-target organisms according to trap type

	No. of organisms (proportion)		
Catch	Box trap	Dome trap (small mesh)	Dome trap (large mesh)
Swimming crab	390 (0.72)	517 (0.85)	76 (0.95)
Other crustaceans	7	20	2
Moray eel	3		
Rockfish	5	5	
Catfish	61	18	
Goby	11	17	
Damselfish	1	4	
Snail	51	23	
Octopus	5	3	
Others	6	2	2
Non-target organisms	150 (0.28)	92 (0.15)	4 (0.05)
No. of traps	100	100	100
Non-target organisms/trap	1.5	0.92	0.04

Dome traps equipped with the smaller mesh netting performed better than box traps by catching more swimming crabs and allowing many non-target organisms to escape through the open funnels. The results of these trials proved that the dome design is superior at capturing swimming crabs and reducing non-target species in the catch and should be used for eradication programs. Box traps caught less swimming crabs than the small meshed dome traps and almost twice as many non-target organisms, making them undesirable as eradication gear. The entrances fitted in the box traps close after an animal pushes through the tight slit acting as a one-way valve, making it impossible for any organism to escape from inside. The catch from dome traps equipped with larger mesh netting (6 cm mesh size) was almost exclusively composed of large commercial swimming crabs, with hardly any non-target organisms. This contrasts with the results obtained with the box traps, which caught many undersized crabs and non-target species. This finding confirms the great selectivity of the dome design for harvesting commercial sized swimming crabs.

After conducting the fishing trials with the different trap designs and determining their entry rates, a main cause for concern was the possible negative effect that lost traps might cause during a large eradication program. Traps lost in a fishing ground continue fishing unattended (ghost fishing) for many years, and this will result in considerable damage to existing fisheries resources. For this reason it was considered necessary to evaluate the escape rates from the traps employed and estimate their possible negative effects.

Escape from Traps

Box and dome trap designs containing crabs were set in a pond and their escape rates were ascertained during a 7-day period (Vazquez Archdale et al. 2007). Individual swimming crabs were marked and placed inside the different traps and escape rates were observed while diving. In several cases octopuses were found to be interfering with some of the traps, as confirmed by their presence together with the crab remains. For this reason complementary observations were conducted in a large tank by placing crabs inside traps, and subjecting them to a 7-day observation period. Escape results showed similar trends in both the pond and tank; crabs could not escape from box traps, as a consequence of their inability to separate the netting panels forming the slit entrance to exit the trap. On the other hand, the open entrances of the dome trap permitted escape easily, and this was more pronounced in the pond than in the tank, where crab recluses were subjected to more outside stimuli. Average residence times for the 7-day observation period were 6.33 days for *P. pelagicus* and 6.83 for *C. japonica* inside the box trap, but only 0.30 days and 2.61 days in the dome trap respectively. Results obtained during the tank experiment using *C. japonica* were similar, with full 7-day residence in the box trap, but only averaging 4.8 days in the dome trap.

From these findings it was concluded that a box trap lost at sea would continue to fish for up to 10 years, which is the average lifetime of a trap according to fishers, while the animals caught inside become bait and attract new ones into the trap. On the other hand, the dome design has open funnel entrances, and will allow crabs and other organisms to escape through them after the original bait is consumed. Consequently, for eradication purposes the dome trap proves to be the best design; it catches the most swimming crabs and, if lost at sea it will not harm so many organisms by ghost fishing and will conserve the existing fisheries resources.

Conclusion

There is still considerable room for improving current baiting practices. By applying the findings of chemoreception research, the effectiveness of conventional fish baits could be almost doubled with the simple addition of small sections of sugarcane. The selectivity potential of the bait was also demonstrated by adding sugar to the fish mince teabags, which maintained the catches of *P. pelagicus* but reduced those of other crab species. In the future, determining the taste preferences of the target organisms can help design species-specific bait, and this will have serious applications in eradication programs. This new "designer" bait would considerably reduce the number of non-target organisms in the catch, and the handling damage and stress they are subjected to while they are in the traps.

Live decoys seem to be an attractive alternative to common fish bait because they only attract crab conspecifics and greatly reduce the non-target catch. In the past pheromone baited traps have been used for the removal of insect pests, and such technology may also be applied in the future to eradicate invasive crustacean species. Recent work on crab pheromones and their isolation might be used to develop pheromone bait by binding these substances in a gel matrix that can be used instead of live decoys (Behrens Yamada et al. 2006).

Trap design profoundly affected the number of crabs found in the catch. Round or oval based traps enhanced crab exploration around their perimeter, and consequently increased the probability of crabs encountering an entrance that would lead them to the bait found inside. Entrance design was crucial as well; tight slit entrances prevented escape but at the same time hindered entry into the traps. The netting material forming the slits tangled with the spines on the carapace and appendages of the crabs, and restricted their forward movement into the traps. Open funnel entrances performed much better because they do not obstruct crab passage and facilitated access towards the bait. This entrance type allowed for easy entry but also permitted escape. Escape has been blamed for decreasing a trap's performance, but recently, as the negative consequences of lost traps by ghost fishing have been documented, the need for escape mechanisms has been recognized. Traps containing sufficient bait quantity will keep the catch inside satisfied until they are hauled, and escape will not be a problem until a considerable time has passed.

The dome trap with open funnel entrances proved to be the best design for both commercial exploitation and for eradication purposes. This trap is ideal for removing crabs of all sizes when fitted with smaller mesh netting to also retain smaller ones. In the future, other ways to improve its design might be to increase its size and volume, so that more space is available and larger crabs do not exclude smaller ones from entering the trap. Increasing the number of entrances would also facilitate entry and increase the probability of the bait odor trail coinciding with one of the entrances and thus effectively guide the crabs into the trap. In extreme cases, fitting the entrances with escape preventing devices could be applied to improve crab retention; soft-eyed entrances or triggers may be installed. In large-scale eradication programs the loss of some trapping gear is inevitable, for this reason fitting all traps with "ghost fishing" prevention biodegradable panels that disintegrate with time, or installing time releases that open some part to make the trap non-operational should be mandatory.

Investigating the possible application of alternative fishing gear like liftnets or tangle nets should be considered for eradicating crustacean pests. In several countries the use of this gear is considered "illegal" and is favored by poachers because they are operated quickly and are very effective. Crustacean fisheries employing tangle nets are found in some countries, and a good example is the spanner crab *Ranina ranina* fishery in Australia, where this gear has proved much more efficient than using conventional traps (Kennelly and Craig 1989). Preliminary trials using liftnets to capture swimming crabs confirmed their superiority to traps, and caught five times more crabs in one hour than dome traps in one day. Liftnets are fished much shorter times, from 30-60 minutes and consume half the amount of bait. When considering the expense of bait, lift nets caught 20 crabs per kilogram of bait, while traps caught only 2.8 (Vazquez Archdale, unpublished results). Liftnets can also be deployed from land, small dinghies and even canoes. They are easily constructed and the materials employed are cheap, so sport fishers and volunteers can easily join eradication campaigns. Funds and manpower to educate the public should be given priority, involving the local community and fishing cooperatives on fishing campaigns and even teaching them how to prepare dishes, consume and market these novel delicious crabs. Even restaurants serving "alien crabs" on their menu may be surprised at the demand their dishes may have among educated customers, who are willing to contribute their part to help reduce the numbers of these invasive pests.

ACKNOWLEDGMENTS

The author is grateful to Mrs. Rebecca Finnis for her valuable time and great improvements on the original manuscript, and Dr. Gunzo Kawamura, from the Faculty of Fisheries, Kagoshima University, for his constant support.

REFERENCES

Ache, B.W. Chemoreception and thermoreception. In: Atwood, H.L. & Sandeman, D.C. *The biology of crustacea (vol 3)*. New York: Academic Press; 1982; 369-398.

Archdale, M.V. & Nakamura, K. (1992). Responses of the swimming crab *Portunus pelagicus* to amino acids and mono- disaccharides. *Nippon Suisan Gakkaishi*, *58,* 165.

Anraku, K., Vazquez Archdale, M., Mendez Cortes, B. & Espinosa R.A. (2001). Crab trap fisheries: capture process and an attempt on bait improvement. *University of the Philippines in the Visayas Journal of Natural Sci*ences, *6,* 121-129.

Burdick, A. Out of Eden: An odyssey of ecological invasion. New York: Farrar, Strauss and Giroux; 2005.

Behrens Yamada, S., Dawkins, M., Quaintance, A., Sugg, L., Bublitz, R. & Hardege, J. (2006). Green Crab Sex Pheromones: New tools for controlling a global invader? *Aquatic Nuisance Species Project Report*, Pacific States Marine Fisheries Commission, Portland, Oregon, 13pp.

Carr, W.S.E. The molecular nature of chemical stimuli in the aquatic environment. In: Atema, J., Fray, R.R., Popper, A.N. & Travolga, W.N. *Sensory biology of aquatic animals*. New York: Sprinter-Verlag Inc.; 1988, 3-27.

Furevik, D.M. Behavior of fish in relation to pots. In: Ferno, A. & Olsen, S. *Marine fish behavior in Capture and abundance estimation*. London: Fishing News Books; 1994; 28-44.

Hardege, J.D., Jennings, A., Hayden, D., Muller, C.T., Pascoe, D., Bentley, M.G. & Clare, A.S. (2002). Novel behavioural assay and partial purification of a female-derived sex pheromone in *Carcinus maenas*. *Marine Ecology Progress Series*, *244,* 179-189.

Heatwole, D.W., Hunt, J.H. & Kennedy, F.S. (1988). Catch efficiencies of live lobster decoys and other attractants in the Florida (USA) spiny lobster fishery. *Florida Marine Research Publications*, *44,*1-15.

Kawamura, G., Matsuoka, T., Tajiri, T., Nishida, M. & Hayashi, M. (1995). Effectiveness of sugarcane-fish combination as bait in trapping swimming crabs. *Fisheries Research*, *22,* 155-160.

Kennelly, S.J. & Craig, J.R. (1989). Effects of trap design, independence of traps and bait on sampling populations of spanner crabs *Ranina ranina*. *Marine Ecology Progress Series*, *51,* 49-56.

Kim, D. & Ko, K. Fishing mechanisms of pots and their modification. 2. Behavior of crab *Charybdis japonica*, to net pots. (1987). *Bulletin of the Korean Society of Fisheries Technology*, *20,* 348-354.

Lafferty, K.D. & Kuris, A.M. (1996). Biological control of marine pests. *Ecology*, *77,* 1989-2000.

Van Engel, W.A. (1958). The blue crab and its fishery in Chesapeake Bay. Part 2. Types of gear for hard crab fishing. *Virginia Institute of Marine Science, 79,* 43-55.

Vazquez Archdale, M., Anraku, K., Yamamoto, T. & Higashitani, N. (2003). Behavior of the Japanese rock crab "Ishigani" *Charybdis japonica* towards two collapsible baited pots: Evaluation of capture effectiveness. *Fisheries Science, 69,* 785-791.

Vazquez Archdale, M.F. & Kuwahara, O. (2005). Comparative fishing trials for *Charybdis japonica* (A. Milne Edwards) using collapsible box-shaped and dome-shaped pots. *Fisheries Science, 71,* 1229-1235.

Vazquez Archdale, M, Kariyazono, L. & Añasco, C.P. (2006a). The effect of two pot types on entrance rate and entrance behavior of the invasive Japanese swimming crab *Charybdis japonica. Fisheries Research 77,* 271-274.

Vazquez Archdale, M., Añasco, C.P. & Hiromori, S. (2006b). Comparative fishing trials for invasive swimming crabs *Charybdis japonica* and *Portunus pelagicus* using collapsible pots. *Fisheries Research, 82,* 50-55.

Vazquez Archdale, M., Añasco, C.P., Kawamura, Y. & Tomiki, S. (2007). Effect of two collapsible pot designs on escape rate and behavior of the invasive swimming crabs *Charybdis japonica* and *Portunus pelagicus. Fisheries Research, 85,* 202-209.

Vazquez Archdale, M., Añasco, C.P. & Tahara, Y. (2008). Catches of swimming crabs using fish mince in "teabags" compared to conventional fish baits in collapsible pots. *Fisheries Research, 91,* 291-298.

In: Invasive Species: Detection, Impact and Control
Editors: C.P. Wilcox and R.B. Turpin
ISBN 978-1-60692-252-1

Chapter 9

"SPECIES POLLUTION" IN FLORIDA: A CROSS-SECTION OF INVASIVE VERTEBRATE ISSUES AND MANAGEMENT RESPONSES

Richard Engeman[1,*], Bernice Constantin[2], Scott Hardin[3], Henry Smith[4,5,6] and Walter E. Meshaka, Jr.[7]

[1]USDA/APHIS/Wildlife Services/National Wildlife Research Ctr, 4101 LaPorte Ave., Fort Collins, Colorado 80521-2154, USA
[2]USDA/ APHIS/Wildlife Services, 2820 East University Ave., Gainesville, Florida 32641, USA
[3]Florida Fish and Wildlife Conservation Commission, 620 South Meridian Street, Tallahassee, Florida 32399, USA
[4]Florida Department of Environmental Protection, Florida Park Service, 13798 S.E. Federal Highway, Hobe Sound Florida 33455, USA
[5]Florida Atlantic University, Wilkes Honors College, 5353 Parkside Drive SR 232, Jupiter, Florida, 33458, USA
[6] (Deceased)
[7]The State Museum of Pennsylvania, 300 North Street, Harrisburg, Pennsylvania 17120, USA

ABSTRACT

The state of Florida has among the two worst invasive species problems in the USA. Besides the sheer numbers of established exotic species in Florida, many present novel difficulties for management, or have other characteristics making effective management extremely challenging. Moreover, initiation of management action requires more than recognition by experts that a potentially harmful species has become established. It also

---

* Correspondence: Richard Engeman Tel: (970) 266-6091 Fax: (970) 266-6089 email: richard.m.engeman@aphis.usda.gov

requires the political will along with concomitant resources and appropriate personnel to develop effective methods and apply them. We illustrate various aspects of the situation in Florida with examples of invasive vertebrates, the problems they pose(d), and management approaches to the problems.

The problems described include long-established widespread and destructive species requiring intensive localized management (feral swine, feral cats); recently established species with potentially severe repercussions, but no broad operational removal programs yet in place (Nile monitor lizards, Burmese pythons), highly prolific mammals that could rapidly invade wide areas without containment/eradication (Gambian giant pouched rats, black-tailed jackrabbits); recently established, potentially destructive birds that might still be eradicated (purple swamp hens); species where sufficient public outcry resulted in control programs (black spiny-tailed iguanas); and rapidly expanding aggressive species for which no practical management actions are available (northern curlytail lizard). A species subset is used here to exemplify in more detail the array of invasive vertebrate species situations in Florida, including routes of introduction, impacts, surrounding politics, and management actions. These examples not only demonstrate the breadth of the terrestrial invasive vertebrate problems in the state, but they also show the diversity in resolve and response among the many species and the motivating factors.

INTRODUCTION

The negative impacts inflicted by exotic species on native species and ecosystems may only be exceeded by human-caused habitat destruction (Parker et al. 1999; Wilcove et al. 1998). In the USA, exotic species have played a role in the listing of 42% of the species protected by the Endangered Species Act (Stein and Flack 1996). Invasive species can be considered "pathogens of globalization" (Bright 1999) and Florida provides an ideal medium in which such pathogens can incubate. In fact, quantitative indicators for assessing non-native species situations are analogous to epidemiological descriptors of disease status in a population (Meyerson et al. 2008). Florida's subtropical climate, its major ports of entry for many wildlife species to the U.S. (both legal and illegal), its thriving $300 million captive wildlife industry, and its position in an area of destructive hurricanes that can release captive animals make the state especially susceptible to the introduction and establishment of a wide range of species (e.g. Corn et al. 2002, Hardin 2007). Moreover, Florida is isolated from land with similar climates, resulting in the state's native vertebrates typically originating in the southeast U.S. at the southern extremes of their range. Invaders to Florida therefore find relatively fewer native species to contend with than in most tropical/subtropical locations (Hardin 2007). Florida joins Hawaii as the two states with the most severe invasive species problems in the United States (U.S. Congress 1993, Corn et al. 2002). Notably, Florida has more introduced animals than any other region of the U.S. and also ranks high in this respect globally, with breeding populations of new vertebrate species regularly identified (SFWMD 2008).

The impacts from many introductions are unknown or not readily perceived by the public, while others are immediately apparent or have their negative potential revealed over time. Even highly prolific invasive species may fester for a considerable time before exhibiting an explosive expansion of their range (Shigesada and Kawasaki 1997). Management of an exotic species requires more than the recognition of a potential problem, it also requires a governmental/public motivation to address the problem. Invasive species often present novel control situations for managers, requiring the acquisition of biological knowledge and the development and testing of control technologies and strategies (see, for example, Engeman and Vice 2001).

The situation in Florida is best understood through a variety of examples. Given Florida's climate, it is no coincidence that a large proportion of the species discussed here are reptiles. Overall, the examples not only demonstrate the breadth of the terrestrial invasive vertebrate problems in the state, but they also show the diversity in resolve and response against the many species and the motivating factors.

FERAL SWINE

Who let the hogs out? In Florida, it originally was the Spanish explorer Hernando de Soto about a half millennium ago in 1539 who introduced feral swine (*Sus scrofa*) to Florida (Towne and Wentworth 1950), and many additional introductions have followed since. This species has the greatest reproductive potential of all free-ranging large mammals in the United States (Wood and Barrett 1979; Hellgren 1999), which combined with a general absence of large predators over much of their range results in continued population increases and range expansions. Swine are well-known for their depredations on crops, livestock and wildlife (e.g., Choquenot 1996; Seward et al. 2004; USDA 2002). In addition, feral swine in Florida have been documented to harbor as many as 45 parasites and infectious diseases (Forrester 1991).

Feral swine are a recreational game animal in Florida, and consequently would not be targeted for eradication (even if that was possible). Furthermore, some claim they are a vital food source for the highly endangered Florida panther (Maehr et al. 1990) (*Felis concolor coryi*). Conversely, feral swine are also a threat to the Florida panther through transmission of pseudorabies virus, as prey-to-panther transmission has been documented to result in the death of the panther (Glass et al. 1994).

The negative environmental impacts of feral swine often require intensive local control. A premium is placed on sanctuaries for protection and preservation of habitats and species in Florida, especially because much of the natural habitat in Florida has been lost to development. There is an ongoing battle in many parts of the state to protect rare habitats from swine damage (e.g., Florida Natural Areas Inventory 1990). Feral swine in Florida have contributed to the decline of at least 22 plant species and 4 species of amphibians listed as rare, threatened, endangered, or of special concern (USDA 2002). Control efforts typically concentrate on conserving special habitats or species, especially in parks and refuges.

Considerable applied research in Florida has been directed towards development of practical in-field methods for implementing, enhancing and evaluating swine removal for resource protection (Engeman et al. 2007b). Methods have been developed for characterizing swine distribution and relative abundance (Engeman et al. 2001, 2007c), and for assessing damage levels in a variety of habitats (Engeman et al. 2003, 2004b, 2007c). An important complement to estimating damage levels was development of credible means to monetarily value their environmental damage (Engeman et al. 2004a). The ability to place monetary values on damage allows the results of management actions to be evaluated in the same metric (dollars) as management expenses. Universally, economic analyses have shown the benefits for swine removal to be remarkable compared to management costs (Engeman et al. 2003; 2004b; 2007c), and to also supersede habitat conservation benefits derived from hunting (Engeman et al. 2007c).

The ability to value the habitat resource provides an effectual economic management tool for evaluating conservation approaches. Economic analyses can greatly assist managers on how to most efficiently and effectively allocate limited funds towards habitat conservation. Ultimately, many conservation funding decisions are made on a political level by people without high levels of training or expertise in biological sciences. While it is essential to obtain high-quality data to understand the biological impacts of management efforts, placing conservation issues in an economic context can greatly enlighten the political decision making process on swine removal and has been a driving force for expanding conservation efforts through swine removal (Engeman et al. 2007b).

NILE MONITOR LIZARDS

Many exotic arrivals to Florida do not appear in the public conscientiousness. For example, the mainstream public is typically unaware that the number of non-native lizard species breeding in Florida now exceeds the number of native species, with over three times as many exotic species as native in south Florida (Hardin 2007), and many of the exotic lizard species can eat various life stages of other lizards (Meshaka et al. 2004). Nonetheless, problems with several large lizard species recently have received public/media attention, a factor sometimes serving to catalyze action. Notable among these are problems from a very large (up to 2.3 m), visible lizard, the Nile monitor (*Varanus niloticus*), which over the last 15+ years has become firmly established in the Cape Coral area (Enge et al. 2004), and also now appears established in the Homestead area (USDA/Wildlife Services unpublished data). Nile monitors have been commonly sold in the U.S. pet trade (Bayless 1991; Faust 2001), although the size and disposition of the adults makes them ill-suited to captivity (Bennett 1995). This species may be on the cusp of no-return in terms of its potential for eradication from Florida. Its range around Cape Coral is expanding into neighboring wildlands, and it also has become established on nearby Pine Island, and possibly Sanibel Island as well, where it would be a threat to endangered sea turtles and shore birds (Enge et al. 2004; Campbell 2005).

The Nile monitor can rapidly outgrow many, if not most, potential predators (Meshaka 2006), and this large-bodied carnivore is capable of eating a wide variety of vertebrate prey, potentially impacting a number of threatened and endangered species in the process (Meshaka 2006). For example, the burrowing owl (*Athene cunicularia*), a Florida Species of Concern, has already been observed as a prey item (Hardin 2007). This is a prolific species capable of reaching high densities (Western 1974). Based on its native range, this lizard could expand its range and pose severe threats to native fauna throughout Florida, and possibly beyond (Enge et al. 2004).

An intense and prompt eradication effort might still eliminate the Nile monitor from Florida. Accumulation of useful information for the management of the species has begun (Campbell 2005). However, this would be a novel species to subject to control activities. Considerable development of methods and technologies would be needed for the implementation of a practical, broad-based control or eradication program. Basic information on diet, baits, and trapping technology exists (Campbell 2005). Considerable testing and refinement of additional baits, attractants, and capture methods applicable to large-scale removal are needed. Building on the successful development of acetaminophen as a toxicant for brown tree snakes (Savarie et al. 2001), trials have been initiated and shown promise for this compound to also be effective for Nile monitors (R. Mauldin and P. Savarie, National Wildlife Research Center unpublished data). Despite a reasonably high profile and media attention, funding has not yet materialized for general development of the needed control technologies, nor for initiating a general control or eradication effort. Without prompt action, the likelihood for successful eradication diminishes. It remains to be seen if denial of "de Nile monitor" will take place in time.

Burmese Pythons

The Burmese python (*Python molurus bivittatus*) is another large exotic carnivore entrenched in Florida (Meshaka et al. 2000). The pathway to invasion for this species has been largely attributed to (illegal) pet releases (e.g., Snow et al. 2006). However, the highly destructive impacts from Hurricane Andrew in 1992 included the release of many animals from captive breeding and holding facilities. Recent genetic results showing little differentiation among pythons captured in south Florida are congruent with this possibility for precipitating the population and the resultant numbers currently observed (Collins et al. 2008).

Similar to the Nile Monitor, there is a diminishing probability for successful eradication as time passes without intensive management action. Its range has been expanding, although the total extent of its potential range in the U.S. has been the subject of considerable controversy (Barker and Barker 2008, Pyron et al. 2008). Nevertheless, containment to its current range may not remain realistic without developing and broadly implementing control methodologies. This very large snake (up to 7 m) has been found with increasing frequency in and around Everglades National Park on the southern tip of Florida. The possibility that this snake might replace the American alligator (*Alligator mississippiensis*) as the top-order carnivore in its range cannot be discounted. In addition, this is one of the six largest snakes in the world, and a large python could pose a danger to humans, especially in Everglades National Park which has over a million visitors annually.

Controlling Burmese pythons in everglades habitats of wet sawgrass prairies with interspersed hardwood hammocks will be challenging. The snake appears vulnerable to approaches that take advantage of its reproductive behaviors. Telemetry trials have already demonstrated on a small scale that female snakes during breeding season can be used as lures to locate males, and telemetered males can be used to locate females (Snow et al. 2006). Since it takes three to five years for Burmese pythons to reach sexual maturity, control based on reproductive behaviors would be a multi-year endeavor to capture animals as they reach sexual maturity.

A set of control tools and strategies were successfully developed for another destructive invasive snake, the brown tree snake on Guam (Engeman and Vice 2001). While the Burmese python is a significantly different species than the brown tree snake, the same conceptual approaches for developing an integrated pest management program can be applied. For example on small scales, multi-capture traps are being designed and research is being conducted on potential attractants within multiple agencies. Similarly, tests also have been initiated into the toxicity to Burmese pythons of acetaminophen, again with promising results (R. Mauldin and P. Savarie, National Wildlife Research Center unpublished data). In Florida, bait placement would need to be specific to Burmese pythons to avoid harming nontarget species. The unique combination of the python's size, dietary potential, and movement ability could be used to make bait delivery specific to the pythons.

The research into control methods and strategies for Burmese pythons has received very limited funding to date, but the technical expertise for developing and implementing control methods is in place should sufficient funding become available for initiating a concerted control effort. Hopefully, the snake's increasingly high profile in the media and in political circles will lead to improved funding in the near future. In the meantime, the range of the snake continues to expand.

Northern Curlytail Lizards

The northern curlytail lizard (*Leiocephalus carinatus armouri*) is endemic to the islands of the Little Bahama Bank, with other subspecies found in the Great Bahama Bank, Cayman Islands, and Cuba (Schwartz and Thomas 1975, Schwartz and Henderson 1991). A small colony established in Palm Beach County through the intentional release of 20 pairs in the 1940s has spread widely (Duellman and Schwartz 1958). Prior to 1968, the range for this population had been expanding north and south along the Atlantic coast at an average rate of 0.98 km/yr, but from 1968 to 2002 it expanded at a much greater average rate of 2.4 km/yr (Smith et al. 2004; Smith and Engeman 2004), and is continuing to expand. Moreover, curlytail lizards are also found in disparate parts of south Florida through human translocations (e.g., Meshaka et al. 2005).

The primary concern with this species' (rapid) range expansion is its depredations on other (small) lizards (Meshaka et al. 2005). Saurophagy is a component of the northern curlytail's ecology (e.g., Smith and Engeman 2004, Dean et al. 2005), and the widely-distributed, also exotic, brown anole (*Anolis segrei*) is a known prey species that could provide expanding populations with a nutritious prey base and a simultaneous reduction in competitors (Meshaka et al. 2005). The northern curlytail is aggressive towards fauna in its size class and was even observed to attack a juvenile northern mockingbird (*Mimus polyglottos*) (Smith and Engeman 2007), the adults of which prey on northern curlytails (Smith et al. 2006). This potential displacer/replacer for the brown anole likely will put the native lizard fauna with which the northern curlytail exists at risk, including state-listed species (Meshaka et al. 2005). The negative impacts would be especially critical in human-disturbed habitat where the northern curlytail lizard is expanding its range and native lizards might already be marginalized.

Although the northern curlytail is unlikely to receive much attention outside herpetological circles, it was described in one newspaper article as “the T-rex of ground critters" (Fleshler 2006). Nevertheless, the northern curlytail lizard, like many of Florida's small-to-medium sized invasive lizards, is unlikely to be targeted for control or eradication.

Its ubiquity within its extended range, small size, and the difficulty in isolating it for control in the presence of native lizard species would make control or eradication difficult, prohibitively expensive, and without the high profile that would engender public support.

BLACK SPINY-TAILED IGUANAS (CTENOSAURS) – "OFF TO SEIZE THE LIZARD"

The black spiny-tailed iguana (*Ctenosaura similis*) on Gasparilla Island is an example of an exotic lizard species (Meshaka et al. 2004) where control was initiated at the behest of affected residents. Also known as ctenosaurs, these large lizards became established on this 11 km-long barrier island along Florida's west coast with an introduction of as few as three individuals around 30 - 35 years ago (Krysko et al. 2003). Since then, the ctenosaur population has saturated the terrestrial habitats on the island in high numbers, including all residential and commercial areas. The boundary line between two counties runs across Gasparilla Island, and the iguanas had become such a nuisance to property owners through damage to landscape plants and homes (especially attics) that residents of both counties voted to self-tax to secure funds for ctenosaur control programs. Moreover, as has been examined for green iguanas (*Iguana iguana*), ctenosaur burrows could undermine public works, such as seawalls and levees, weakening them for withstanding severe storm events (Sementelli et al. 2008).

Ctenosaurs conflict with a variety of ecological interests in addition to the economic interests on the island. While Gasparilla Island is largely developed, it also is the location for Gasparilla Island State Park, 49 ha of mostly natural area on the southern end of the island (FDEP 2002). Also despite the development, Gasparilla Island's beaches are home or potential nesting site for a variety of species federally or state-listed as threatened, endangered or of concern (FDEP 2002). The endemic listed species on Gasparilla Island for which this species may pose a threat include eggs and young of nesting shorebirds, beach mice, hatchling sea turtles and gopher tortoises (*Gopherus polyphemus*) (Krysko et al. 2003). It may also pose a threat to attack snakes on the island (Engeman et al. in press), including some size classes of eastern indigo snakes (*Drymarchon corais couperi*), a threatened species (Moler 1992). Further environmental impacts include a mutualistic association between ctenosaurs and Brazilian pepper (*Schinus terebinthifolius*), the most problematic invasive plant on Gasparilla Island (FDEP 2002, Jackson and Jackson 2007). Populations of both species are enhanced by ctenosaur foraging on Brazilian pepper (Jackson and Jackson 2007). Invasive plant control is time consuming and costly, and the ctenosaur serves to increase the problem and raise potential remediation costs.

Active iguana removal was implemented in both counties to reduce, and ultimately eradicate if possible, their populations, albeit differing approaches have been applied in the two counties. Lee County on the southern portion of the island applied a sole-source bounty system whereby a reward has been paid to a contractor for each lizard removed (by a variety of methods). Charlotte County on the northern portion of the island formed an agreement to remove ctenosaurs with the U.S. Department of Agriculture's Wildlife Services (WS), the federal agency authorized to resolve human-wildlife conflicts. Their multi-faceted approach includes population monitoring, iguana removal (also by a variety of methods), and research to develop and evaluate control methods (including toxicant screening tests). Over time, the approaches by the two counties will provide an interesting comparison in efficacies and economics. The cost-per-lizard to remove iguanas in Lee County remains constant, whereas the cost-per-lizard decreases with each subsequent iguana removed in Charlotte County. Once the number of ctenosaurs captured in Charlotte County exceeds the amount of the agreement divided by the amount of the Lee County bounty, then the Charlotte County approach becomes more cost-effective.

PURPLE SWAMPHEN

Relatively few non-native bird species have become established in Florida (Hardin 2007), as only about 5% of the roughly 200 non-native species introduced have succeeded at becoming established (Avery 2007). The purple swamphen (*Porphyrio porphyrio*), a recent introduction to Florida, was judged to merit eradication by a consensus of land management agencies based on its increasing population and range expansion, its potential impact to native species, and the potential for an eradication effort to succeed (Ferriter et al. 2008, Hardin 2007). This large rail species is native to Europe, Africa, Asia and Australia. It also is native to American Samoa, a factor potentially complicating its control in Florida if eradication efforts were delayed (Ferriter et al. 2008). Because it is native to American Samoa, the purple swamphen is being considered for inclusion to the Migratory Bird Treaty Act (MTBA). However, the MTBA provides protection for a species throughout all U.S. holdings and historically has not made geographic distinctions within the U.S., which could protect purple swamphens from removal in Florida in the future. This factor increased the urgency to move on the Florida population.

The species was first observed in the wild in urban southeast Florida in 1996, where the population resulted from escapes from a local aviculturist or escaped from the Miami Metrozoo in 1992 as a result of Hurricane Andrew (Avery 2007, Ferriter 2008, Hardin 2007). As the population increased to over 200 birds, it still remained only in developed areas, but by 2006 it had expanded its range to Everglades Conservation Areas and has been reported as far north as Lake Okeechobee (Hardin 2007). Efforts to eliminate the purple swamphen were prompted by its ecological similarity to the native common moorhen (*Gallinula chloropus*) and purple gallinule (*Porphyrula martinica*) and the looming potential for it to be protected by the MTBA (Ferriter 2008, Hardin 2007).

Purple swamphen control was initiated in 2006 in a cooperative effort among biologists with the South Florida Water Management District (SFWMD), the U.S. Fish and Wildlife Service (USFWS), and the Florida Fish and Wildlife Conservation Commission (FWC). Over 800 birds were located and removed during October 2006-August 2007 (Clary 2007). Efforts are scheduled to continue to remove the remainder of the introduced population. At the least, potential impact to native wildlife and vegetation can be minimized, or at the best, the species will be eradicated from Florida (Avery 2007, Hardin 2007).

Feral Cats

The FWC estimates there are between 6.3 and 9.6 million feral cats (*Felis catus*) in Florida (at http://www.floridaconservation.org), which, conservatively, kill millions of small animals in Florida each year (FCIT 2003). Feral cats are generally harmful to native fauna throughout the state, because even cats well-maintained as pets take a high toll of nearby small animals (Churcher and Lawton 1987, Lepcyzk et al. 2003, Woods et al. 2003), especially considering cats continue to hunt and kill when not hungry (Liberg 1984). Globally, feral cats feed heavily on small vertebrates and have led to the extinctions of a number of species (e.g., Burbidge and Manly 2002, Nogales et al. 2003). Feral cats in Florida have been observed to prey on loggerhead (*Caretta caretta*) and green (*Chelonia mydas*) sea turtles, roseate tern (*Sterna dougallii*), least tern (*Sterna antillarum*), American oystercatcher (*Haematopus ullietus*), Florida scrub jay (*Aphelocoma coerulescens*), Choctawhatchee beach mouse (*Peromyscus poloionotus allophrys*), Anastasia Island beach mouse (*Peromyscus gossypinus gossypinus*), Key Largo cotton mouse (*Peromyscus gossypinus allapaticola*), Southeastern beach mouse (*Peromyscus poloionotus niveiventris*), Perdido Key beach mouse (*Peromyscus poloionotus trissyllepsis*), Key Largo woodrat (*Neotoma floridana smalli*), Lower Keys marsh rabbit (*Sylvilagus palustris hefneri*), all federally listed as threatened or endangered (FCIT 2003, Ferriter et al. 2008). Cat removal has been demonstrated to result in immediate rebounds of endangered beach mouse populations (FCIT 2003).

While cats are harmful to wildlife throughout Florida, they are of the particular concern on the islands of the Florida Keys (Ferriter et al. 2008). They have been a factor in the 50% decline in populations of the endangered Lower Keys marsh rabbit (Forys and Humphrey 1999) and cat removal was identified as an integral component in the recovery of Key Largo woodrat (USFWS 1999, 2003). Making matters worse, feral cat colonies can concentrate a large number of instinctive predators in an area and pose significant threats to the smaller fauna in the vicinity. For example, the Ocean Reef Cat Club (ORCAT) at the exclusive Ocean Reef Club residential resort on Key Largo maintains a large feral cat colony adjacent to the federal and state lands supporting the Key Largo woodrat and Key Largo cotton mouse. Despite the protected habitat, the Key Largo woodrat population dropped from 6500 in 1988 to less than 80 animals by the early 2000s (Humphrey 1988, Winter 2004, B. Muiznieks pers. comm.). ORCAT runs an intensive, well-funded trap, neuter and release (TNR) program (Clark and Pacin 2002), but TNR programs are not effective for managing feral cat populations under most circumstances (e.g., Anderson et al. 2004, FCIT 2003, Ferriter et al. 2008, Jessup 2004). The ORCAT colony continues to have around 500 cats neighboring endangered species habitat despite the intensive TNR efforts. Luckily, captive breeding is now helping replenish the Key Largo woodrat population.

Feral and free-ranging cats are notorious for their destruction of avifauna, and this problem is particularly pronounced in Florida where there are large numbers of cats often in the immediate vicinity of small forest remnants and hammocks that migrant birds rely on as migration stopover sites (Winter and Wallace 2006). For example, severe weather in spring 2001 resulted in a massive fallout of migrating warblers in the Keys, where large numbers of the birds were lost to predation by cats (Winter and Wallace 2006). Similarly, the decline of upland bird populations between 1988 and 1998 at Greynolds Park (Miami-Dade County) was due to a cat colony in the park. The problem was rectified by stricter laws against abandoning or feeding cats, and removal of the existing cats (Winter and Wallace 2006).

Beyond their environmental impacts, feral cats, especially in dense colonies, present human and wildlife disease concerns. While cats can carry a host of diseases and parasites, rabies is the greatest concern. Cats are the most frequently reported domestic animal in the U.S. with rabies (e.g., Barrows 2004). For example, between 1988 and 2003, there were 208 laboratory confirmed diagnoses of cats with rabies in Florida (Barrows 2004). In fact, the Florida Rabies Advisory Committee stated "the concept of managing free-roaming/feral cats is not tenable on public health grounds because of the persistent threat posed to communities from injury and disease" (Barrows 2004, Brooks 1999).

While the threats cats pose to native wildlife, especially endangered species, and the disease concerns are well-documented, removal of feral cats, particularly at cat colonies, is often accompanied by vocal public outcry from cat enthusiasts. The efforts to protect highly endangered species in the Florida Keys from predation by cats are noteworthy examples. These programs did not involve the lethal removal of animals. Rather, feral cat removal involved live trapping and turning cats over to animal shelters. Despite that, opposition to cat removal has been stiff (including sabotage of traps) and has affected management of the endangered species. Thus, feral cats, in particular, can present additional social dimensions creating difficulties for effective population management.

Gambian Giant Pouched Rats

The Gambian giant pouched rat (*Cricetomys gambianus*) in the Florida Keys is an example of how a severe invasive species threat can be managed in a logical, practical and efficient manner, once a threat has been identified. This largest of rat species (up to 2.8 kg; Rosevear 1969) is highly prolific and holds potential for extreme ecological and agricultural negative impacts (Perry et al. 2006, Engeman et al. 2006). Although the species escaped from a captive breeder on Grassy Key around 1999, it was not identified as established in the wild until residents brought it to the attention of the USFWS in 2004. Perry et al. (2006) established the existence of a breeding population and its dispersion potential was subsequently modeled (Peterson et al. 2006). Dispersal of the species to mainland Florida could have resulted in continued spread through much of North America where significant negative ecological and agricultural consequences could ensue (Peterson et al. 2006).

Following verification of the population's existence and confirmation of its invasive and destructive potential, information and methods essential for successful eradication were rapidly developed, including detection and monitoring technologies, population indexing methodologies, population distribution, habitat preferences, trapping methodology, acceptance of bait matrices, efficacy tests of toxicants, and bait stations that exclude native species (Engeman et al. 2006, 2007d). To test and fine-tune the methods prior to implementing full-scale eradication, a pilot eradication project was implemented on Crawl Key, a small key adjoining Grassy Key to which the species expanded its range. Afterwards, surveys found no evidence of Gambian giant pouched rats remaining on Crawl Key, although Hurricane Wilma undoubtedly also contributed to their mortality. The criteria (see Engeman et al. 2006, Parkes and Murphy 2003) were considered obtainable for a successful eradication to commence on larger Grassy Key, location of the much larger primary population. Surveys following the hurricane on Grassy Key verified the survival of the Gambian giant pouched rat population, and with a greater range than previously thought. Next and a little over two years after the initial report, the full-scale operation to eliminate this population occurred. At least two years of monitoring for Gambian giant pouched rats should be applied to both Grassy and Crawl Keys, as well as other potential sites of occupancy such as refuse transfer sites (including the mainland landfills) and locations of credible reports of sightings should also receive continued monitoring to help insure no propagules from Grassy Key are surviving elsewhere.

Thus, the rapid response eradication effort for Gambian giant pouched rats can be described as a progression of accomplishments:

1. Verify presence
2. Develop detection and population monitoring methods
3. Develop and test potential control tools
4. Test eradication approach (Crawl Key)
5. Apply eradication methods and strategies to Grassy Key
6. Surveillance for survivors or satellite populations

The eradication effort is currently in the surveillance phase. This phase appears to be working well, as the Gambian giant pouched rats that have occasionally been detected on Grassy Key have been successfully targeted for removal. No Gambian giant pouched rats have been detected outside of Grassy Key.

Whereas the logic and flow described here for this eradication effort makes it seem as though the path to Gambian giant pouched rat eradication was a smooth continuum once the problem was identified and verified, it was, in reality, a series of fits and starts (Engeman et al. 2007d). No single block of funding was available to develop the necessary information and implement an eradication effort. Funding and in-kind resources were provided from > 10 federal, state, and local government entities, as well as private concerns. Even Hurricane Wilma may have assisted the eradication effort to some degree, as it struck at a time lessened resources were available for the work. The storm surge overwhelmed a large part of many of the keys, possibly removing small propagule populations.

One potential pitfall that hopefully will not occur is complacency at the apparent success so far. That could undermine availability of necessary resources to see the follow-up monitoring through to its conclusion. Lack of continued vigilance could result in the hard work to date being undone, or worse if survivors or propagules go undetected, eventual Gambian giant pouched rat dispersal to the mainland. On the other hand, successful eradication of this species hopefully would help reduce the general reluctance of managers to attempt eradications of other invasive species in Florida (see, for example the comments by Donlan et al. 2003).

BLACK-TAILED JACKRABBITS

Black-tailed jackrabbits (*Lepus californicus*) are not native to Florida, but by 2003 they had been well-established at Miami International Airport (MIA) for many years. How and when they were introduced to this expansive airport property was unknown. Speculations as to their origins included escapes from a rabbit farm or escapes from transit to dog racing tracks for use in training greyhounds. By 2003, the black-tailed jackrabbit population at MIA was considered to be around 500. They also had been observed in other parts of Florida, but not as breeding populations.

Occupation of the MIA property by a large number of black-tailed jackrabbits posed two serious threats (see Engeman et al. 2007a). First, even though MIA is relatively encapsulated by the Miami metro area, the jackrabbits still posed a significant invasive threat for Florida. The species is highly fecund, and they also are a highly mobile, fast-moving species. Once outside the confines of Miami they could rapidly spread through Florida (and beyond). The other significant problem their population posed was to cause a severe increase in bird airstrike hazards. Black-tailed jackrabbits were often killed by collisions with aircraft and vehicles, or the back-blast from jet engines. Their carcasses proved highly attractive to vultures (*Cathartes aura* and *Coragyps atratus*) for forage. This created a considerable air safety concern, as vultures present significant hazards to aircraft while taking off or landing (e.g., Dolbeer et al 2000). Besides safety concerns, bird strikes also result in lost revenue and very costly aircraft repairs.

Removal of the black-tailed jackrabbit population at MIA was instigated as a response to Federal Aviation Administration (FAA) regulations mandating the problem be solved for safety reasons (From March 2001 to March 2003 at least two dozen vultures were struck by aircraft at MIA). Thus, the human safety issue motivated their removal, rather than their potential for ecological harm should they have dispersed from MIA. Had it not been for their exacerbation of airstrike hazards at MIA, it is unlikely they would have been eradicated and their population would have continued to be a festering threat for eventual dispersal.

The eradication also revealed the political, economical and social complexities involved in carrying a conceptually straight-forward, but highly visible process. The eradication was delayed multiple times to assuage public sentiment towards lethal control by allowing a live-trapping and translocation (to Texas) attempt to proceed first. That endeavor was unsuccessful at removing more than a portion of the population. Finally, a court ruling allowed the eradication effort to go forward for the sake of the flying public's safety. Lethal removal was efficient and effective at eliminating the black-tailed jackrabbit population, and at a significantly lower cost than the live-trapping venture (Engeman et al. 2007a).

The political, economic and management paths to that success may have been convoluted, but because a human safety concern was clearly recognized, the black-tailed jackrabbit apparently became the first well-established invasive vertebrate species intentionally eradicated from Florida.

DISCUSSION

The invasive species situation in Florida is severe, and when one considers the climatic, demographic, and environment situation in Florida, the severity of the problem is even greater than at first glance. The breadth of invasive land animals in Florida that arguably merit eradication, or at least control, is extensive. The list is extremely varied and includes animals ranging from unusual species of distant origins to more recognizable invaders brought in from other states in the U.S., as well as feral domestics. A variety of steps have been taken to reduce the number of introductions, with some apparent success (Hardin 2007). As is often stated (e.g., NISC 2001), prevention is the most efficient and economical means to eliminate exotic species. However, even if no new exotic vertebrates become established in Florida, there is an abundance of established exotic vertebrates that merit management action.

A brief sampling of Florida's invasive species originating from other parts of the globe (besides those detailed already) include species such as the common boa (*Boa constrictor*), Cuban treefrog (*Osteopilus septentrionalis*), green iguana (*Iguana iguana*), black and white tegu (*Tupinambis merianae*), peafowl (*Pavo cristatus*), spectacled caiman (*Caiman crocodilus*), monk parakeet (*Myiopsitta monachus*), rhesus monkey (*Macaca mulatta*), sacred ibis (*Threskiornis aethiopicus*). Species native to the U.S., but exotic in Florida include animals such as coyotes (*Canis latrans*), red fox (*Vulpes vulpes*), black-tailed prairie dogs (*Cynomys ludovicianus*), red-eared slider (*Trachemys scripta elegans*) and armadillos (*Dasypus novemcinctus*), while other feral domestics widespread in Florida besides swine and cats include goats (*Capra hirus*) and dogs (*Canis familiaris*). Many other less noticeable species have become established in Florida, and other species are suspected to be breeding there, but without firm documentation. Not all will be subjected to eradication or control, but some of the invaders could present potentially severe environmental, human health, and/or economic consequences if their populations are not controlled or eradicated.

Species such as Gambian giant pouched rats, Nile monitors, ctenosaurs, Burmese pythons and many of the other exotic species represent novel species to be considered for eradication or control. The Gambian giant pouched rat, purple swamphen and black-tailed jackrabbit are examples of how necessary incentive and resources can be applied to directly design and implement a practical eradication or control program (Engeman et al. 2006, 2007a). Too often, invasive species merely become the subjects of biological and population studies (e.g., Campbell 2007), but there is a limit at some point to the utility in conducting biological studies of introduced species unless the results directly assist in their removal (e.g., Donlan et al. 2003, Simberloff 2003, Campbell 2007). Donlan et al. (2003) concluded that research directly facilitating eradication tools and projects should be of high priority. Developing the information and technologies from which control strategies can be developed and implemented is an essential component to addressing many invasive species situations. Equally important is the development of public and governmental motivation, i.e. funding, to manage invasive species before their populations expand beyond feasible control.

Many of the problematic invasive vertebrates in Florida are predators. Predation not only threatens many rare species (Hecht and Nickerson, 1999), but the deleterious impacts of predation losses is compounded by habitat loss (Reynolds and Tapper, 1996). Predators also increase the risk of catastrophic extinction of prey populations (Schoener et al. 2001). Given the amount of habitat lost to development in Florida and the state's proclivity for catastrophic hurricanes (two circumstances magnified on the Keys and other islands), a number of species in Florida are at high risk. Since alien predators are more dangerous than native predators to prey populations (Salo et al. 2007), the impacts from invasive predators, whether small like northern curlytail lizards or large like Burmese pythons, could have devastating impacts on Florida's native species, especially the listed rare species.

For a number of well-established species in Florida, such as feral swine, feral cats and green iguanas, there is no practical means to eradicate them from the state. That does not mean they cannot be intensively controlled, managed, or eradicated in situations of greatest priority on a localized scale, especially islands. For example, feral swine are ubiquitous and destructive, but as already discussed, would never be considered for state-wide eradication. However, swine have been successfully targeted in a nearly-completed eradication effort on Cayo Costa and Punta Blanca Islands, with concomitant dramatic improvements in nesting by listed sea turtles and shorebirds. Species like the black-tailed jackrabbit, Gambian giant pouched rat and purple swamphen were identified as feasible, practical, and valuable to eradicate before they become too deeply entrenched across a broad range. To that end, Parkes and Murphy (2003) delineated some "obligate rules" for successful eradication: 1) all individuals of the target species must be at risk of being killed, 2) target species must be removed at a rate greater than the rate they replace their losses, and 3) the risk of immigration must be zero. Given suitable control methods applied in a systematic and sustained integrated pest management program, these criteria could well be met for a number of invasive species in Florida, if their populations are not permitted to fester into an unmanageable situation. The case of the black-tailed jackrabbit demonstrates that, even with many political gyrations, a population of a species with a restricted range can be eradicated without an excessive outlay of resources (Engeman 2007a). To leave such a situation unaddressed is like leaving a slow-burning fuse lit to an ecological bomb.

REFERENCES

Anderson, M.C., Martin, B.J. & Roemer, G.W. (2004). Use of matrix population models to estimate the efficacy of euthanasia versus trap-neuter-return for management of free-roaming cats. *Journal of American Veterinary Medical Association,* 225, 1871-1876.

Avery, M.L. & Moulton, M.P. (2007). Florida's non-native avifauna. In G. Witmer, W. Pitt & K. Fagerstone. (eds.) *Managing Vertebrate Invasive Species: Proceedings of an International Symposium*. Pp 365-377. Fort Collins, CO, USDA/APHIS/WS, National Wildlife Research Center.

Barker, D.G. & Barker, T.M. (2008).Comments on flawed a herpetological paper and an improper and damaging news release from a government agency. *Bulletin of Chicago Herpetological Society*, 43, 45-47.

Barrows, P.L. (2004). Professional, ethical, and legal dilemmas of trap-neuter-release. *Journal of American Veterinary Medical Association*, 225, 1365-1369.

Bayless, M.K. (1991). A trip from Africa. *VaraNews* 1, 5-8.

Bennett, D. (1995). *A little book of monitor lizards*, Glossop, UK, Viper Press.

Bright, C. (1999). Invasive species: pathogens of globalization. *Foreign Policy*, Issue 116, 50-64.

Brooks, R.G. (1999). *1999 rabies prevention and control in Florida.* 1999 Epidemiological Series, Tallahassee, Florida, Florida Department of Health.

Burbidge, A.A. & Manly, B. (2002). Mammal extinctions on Australian islands. *Journal of Biogeography*, 29, 465-473.

Campbell, T.S. (2005). *Eradication of introduced carnivorous lizards from the Cape Coral area*. Final Report to the Charlotte Harbor National Estuary Program, Fort Meyers, Florida.

Campbell, T.S. (2007). The role of early detection and rapid response in thwarting amphibian and reptile introductions in Florida. In G. Witmer, W. Pitt & K. Fagerstone. (eds.) *Managing Vertebrate Invasive Species: Proceedings of an International Symposium*. Pp 146-156. Fort Collins, CO, USDA/APHIS/WS, National Wildlife Research Center.

Choquenot, D., McIlroy, J. & Korn, T. (1996). *Managing Vertebrate Pests: Feral Pigs*. Canberra, ACT, Australia, Bureau of Resource Sciences, Australian Government Publishing Service.

Churcher,P.B. & Lawton, J.H. (1987). Predation by domestic cats in an English village. *Journal of Zoology London*, 212, 439-455.

Clark, A.L. & Pacin, T. (2002). Domestic cat "colonies" in natural areas: a growing exotic species threat. *Natural Areas Journal*, 22, 154-159.

Clary, M. (2007). State aims to eradicate exotic purple swamphens in wetlands. *South Florida Sun-Sentinal*, September 8, 2007.

Collins, T.M., Freeman, B. & Snow, R. (2008*). Genetic characterization of populations of the nonindigenous Burmese python in Everglades National Park*. Final report to South Florida Water Management District.

Corn, M.L., Buck, E.H., Rawson, J., Segarra, A. & Fischer, E. (2002). *Invasive Non-Native Species: Background and Issues for Congress*. Washington, DC, Congressional Research Service and The Library of Congress.

Dean, C.L., Engeman, R.M., Smith, H.T. & Meshaka, W.E.Jr. (2005). Leiocephalus carinatus armouri (northern curly-tailed lizard) Cannibalism. *SSAR Herpetological Review,* 36, 451.

Dolbeer, R.A., Wright, S.E. & Cleary, E.C. (2000). Ranking the hazard level of wildlife species to aviation. *Wildlife Society Bulletin*, 28, 372-378.

Donlan, C.J., Tershy, B.R., Campbell, K. & Cruz, F. (2003). Research for requiems: the need for more collaborative action in eradication of invasive species. *Conservation Biology*, 17, 1850-1851.

Duellman, W.E. & Schwartz, A. (1958). Amphibians and reptiles of southern Florida. *Bulletin Florida State Museum,* 3,181-342.

Enge, K.M., Krysko, K.L., Hankins, K.R., Campbell, T.S. & King, F.W. (2004). Status of the Nile monitor (Varanus niloticus) in southwestern Florida. *Southeastern Naturalist,* 3, 571-582.

Engeman, R.M., Constantin, B.U. & Bunting, J. (2007a). The political, economical and management history of a successful exotic eradication: the case of black-tailed jackrabbits in Florida. In G. Witmer, W. Pitt & K. Fagerstone. (eds.) *Managing Vertebrate Invasive Species: Proceedings of an International Symposium*. Pp 476-478. Fort Collins, CO, USDA/APHIS/WS, National Wildlife Research Center.

Engeman, R.M., Constantin, B., Nelson, M., Woolard, J. & Bourassa, J. (2001). Monitoring changes in feral swine population and spatial distribution of activity. *Environmental Conservation*, 28, 235-240.

Engeman, R.M., Constantin, B., Shwiff, S.A., Smith, H.T., Woolard, J., Allen, J. & Dunlap, J. (2007b). Research in support of feral swine removal efforts in Florida. *Human-Wildlife Conflicts,* 1, 178-185.

Engeman, R.M., Kennedy, M., Constantin, B.U., Christie, M.L. & Hall, P.T. (in press). Ctenosaura similis (black spinytail iguana), Coluber constrictor priapus (southern black racer) nonpredatory killing. *Herpetological Review*.

Engeman, R.M., Shwiff, S.A., Smith, H.T. & Constantin, B.U. (2004a). Monetary valuation of rare species and imperiled habitats as a basis for economically evaluating conservation approaches. *Endangered Species Update,*. 21, 66-73.

Engeman, R.M., Smith, H.T., Severson, R., Severson, M.A., Woolard, J., Shwiff, S.A., Constantin, B.U. & Griffin, D. (2004b). Damage reduction estimates and benefit-cost values for feral swine control from the last remnant of a basin marsh system in Florida. *Environmental Conservation,* 31, 207-211.

Engeman, R.M., Smith, H.T., Shwiff, S.A., Constantin, B.U., Nelson, M., Griffin, D. & Woolard, J. (2003). Prevalence and economic value of feral swine damage to native habitat in three Florida state parks. *Environmental Conservation,* 30, 319-324.

Engeman, R.M, Stevens, A., Allen, J,. Dunlap, J., Daniel, M., Teague, D. & Constantin, B. (2007c). Feral swine management for conservation of an imperiled wetland habitat: Florida's vanishing seepage slopes. *Biological Conservation*, 134, 440-446.

Engeman, R.M. & Vice, D.S. (2001) Objectives and integrated approaches for the control of brown tree snakes. *Integrated Pest Management Reviews*, 6, 59-76.

Engeman, R.M., Woolard, J.W., Perry, N.D., Witmer, G., Hardin, S., Brashears, L., Smith, H.T., Muiznieks, B. & Constantin, B.U. (2006). Rapid assessment for a new invasive species threat: the case of the Gambian giant pouched rat in Florida. *Wildlife Research,* 33, 439-448.

Engeman, R.M., J.W. Woolard, G. Witmer, B.U. Constantin, S. Hardin, N.D. Perry & P. Hall. (2007c). The path to eradication of the Gambian giant pouched rat in Florida. In G. Witmer, W. Pitt & K. Fagerstone. (eds.) *Managing Vertebrate Invasive Species: Proceedings of an International Symposium*. Pp 305 - 311. Fort Collins, CO, USDA/APHIS/WS, National Wildlife Research Center.

Faust, R.J. (2001). Nile monitors: everything about history, care, nutrition, handling, and behavior. Hauppauge, New York, Barron's Educational Series.

FCIT (Feral Cat Issue Team). 2003. *Issue assessment: impacts of feral and free-ranging domestic cats on wildlife in Florida*. Tallahassee, Florida, Florida Fish and Wildlife Conservation Commission.

Ferriter, A.; Doren, B., Thayer, D., Miller, B., Thomas, B., Barrett, M., Pernas, T., Hardin, S., Lane, J., Kobza, M., Schmitz, D., Bodle, M., Toth, L, Rodgers, L., Pratt, P., Snow, S., & Goodyear, C. (2008) South Florida Environmental Report Ch. 9: *The Status of Nonindigenous Species in the South Florida Environment*. West Palm Beach, Florida, South Florida Water Management District.

FDEP (Florida Department of Environmental Protection). (2002) *Gasparilla Island State Park Management Plan*. Tallahassee, Florida, USA, Florida Department of Environmental Protection.

Fleshler, D. (2006). The T-rex of ground critters. *South Florida Sun-Sentinel*, May 1, 2006, 1A-2A.

Florida Natural Areas Inventory (FNAI). (1990). *Guide to the natural communities of Florida*. Tallahassee, Florida, Florida Natural Areas Inventory and Florida Department of Natural Resources.

Forrester, D.J. (1991). *Parasites and diseases of wild mammals in Florida*. Gainesville, Florida, University of Florida Press.

Forys, E.A. & Humphrey, S.R. (1999). Use of population viability analysis to evaluate management options for the endangered Lower Keys marsh rabbit. *Journal of Wildlife Management*, 63, 160-251.

Glass, C.M., McLean, R.G., Katz, J.B., Maehr, D.S., Cropp, B.C., Kirk, L.J., McKeirman, A.J. & Evermann, J.F. (1994). Isolation of pseudorabies virus (Aujeszky's disease) from a Florida panther. *Journal of Wildlife Diseases*, 30, 180-184.

Hardin, S. (2007). Managing non-native wildlife in Florida: state perspective, policy and practice. In G. Witmer, W. Pitt & K. Fagerstone. (eds.) *Managing Vertebrate Invasive Species: Proceedings of an International Symposium*. Pp 43- 52. Fort Collins, CO, USDA/APHIS/WS, National Wildlife Research Center.

Hecht, A., & Nickerson, P.R. (1999). The need for predator management in conservation of some vulnerable species. *Endangered Species Update,* 16, 114-118.

Hellgren, E. (1999). Reproduction in feral swine. In *Proceedings of the 1999 National Feral Swine Symposium*. Pp 67-68. Austin, Texas, Texas Animal Health Commission.

Humphrey, S.R. (1988). Density estimates of the endangered Key Largo woodrat and cotton mouse (*Neotoma floridana smalli* and *Peromyscus gossypinus allapaticola*) using the nested grid approach. *Journal of Mammalogy*, 69, 524-531).

Jackson, J.A. & Jackson, B.J.S. (2007). An apparent mutualistic association between invasive exotics: Brazilian pepper (Schinus terebinthifolius) and black spiny-tailed iguanas (Ctenosaura similis). *Natural Areas Journal*, 27, 254-257.

Jessup, D.A. (2004). The welfare of feral cats and wildlife. *Journal of American Veterinary Medical Association*, 225, 1377-1383.

Krysko, K.L., King, F.W., Enge, K.M. & Reppas, A.T. (2003). Distribution of the introduced black spiny-tailed iguana (Ctenosaura similis) on the southwestern coast of Florida. *Florida Scientist*, 66, 74-79.

Lepczyk, C.A., Mertig, A.G. and Liu, J. (2003). Landowners and cat predation across rural-to-urban landscapes. *Biological Conservation,* 115, 191-201.

Liberg, O. (1984). Food habits and prey impact by feral and house-based domestic cats in a rural area of southern Sweden. *Journal of Mammalogy*, 65, 424-432.

Maehr, D.S., Belden, R.C., Land, E.D. & Wilkins, L. (1990). Food habits of panthers in southwest Florida. *Journal of Wildlife Management*, 54, 420-423.

Meshaka, W.E. (2006). An update on the list of Florida's exotic amphibians and reptile species. *Journal of Kansas Herpetology,* 19, 16-17.

Meshaka, W.E., Butterfield, B.P. & Hauge, J.B. (2004). *The exotic amphibians and reptiles of Florida*. Malabar, Florida, Krieger Publishing Company.

Meshaka, W. E., Jr., Loftus, W.F. & Steiner, T. (2000). The herpetofauna of Everglades National Park. *Florida Scientist,* 63, 84-103.

Meshaka, W.E., Smith, H.T., Engeman, R.M., Dean, C.L., Moore, J.A. & O'Brien, W.E. (2005). The geographically contiguous and expanding range of the northern curlytail lizard (Leicephalus carinatus armouri) in Florida. *Southeastern Naturalist,*. 4, 521-526.

Meyerson, L.A., Engeman, R. & O'Malley, R. (2008) Tracking non-native vertebrate species: indicator design for the United States. *Wildlife Research,* 35, 235-241.

Moler, P.E. (1992). *Rare and endangered biota of Florida, Vol III, Amphibians and Reptiles*. Gainesville, Florida, University Press of Florida.

NISC (National Invasive Species Council). (2001). *Meeting the invasive species challenge: National Invasive Species Management Plan*. Washington, DC, National Invasive Species Council.

Nogales, M., Martín, A., Tershy, B.R., Donlan, C.J., Veitch, D., Puerta, N., Wood, B. & Alonso, J. (2003). A review of cat eradication on islands. *Conservation Biology*, 18, 310-319.

Parker, I.M., Simberloff, D., Lonsdale, W.M., Goodell, K., Wonham, M., Kareiva, P.M., Williamson, M.H., Holle, B.V., Moyle, P.B., Byers, J.E. & Goldwasser, L. (1999).

Impact: toward a framework for understanding the ecological effects of invaders. *Biological Invasions,* 1, 3-19.

Parkes, J., & Murphy, E. (2003). Management of introduced mammals in New Zealand. *New Zealand Journal of Zoology,* 30, 335-359.

Perry, N.D., Hanson, B., Hobgood, W., Lopez, R.L., Okraska, C.R., Karem, K., Damon, I.K., Carroll, D.S. (2006). New invasive species in southern Florida: Gambian rat (*Cricetomys gambianus*). *Journal of Mammalogy,* 87, 262-264.

Peterson, A.T., Papes, M., Reynolds, M., Perry, N.D., Hanson, B., Regnery, R., Hutson, C., Muizniek, B., Damon, I. & Carroll, D. (2006). Native range ecology and invasive potential of *Cricetomys* rats in North America. *Journal of Mammalogy,* 87, 427-432.

Pyron, R.A., Burbrink, F.T. & Guiher, T.J. (2008). Claims of potential expansion throughout the U.S. by invasive python species are contradicted by ecological niche models. *PLoS ONE*, 3(8): e2931.

Reynolds, J.C., & Tapper, S.C. (1996). Control of mammalian predators in game management and conservation. *Mammal Review,* 26, 127-156.

Rosevear, D.R. (1969). 'he Rodents of West Africa. London, British Museum of Natural History.

Salo, P., Korpimäki, E., Banks, P. B., Nordström, M. & Dickman, C. R. (2007). Alien predators are more dangerous than native predators to prey populations. *Proc. Royal Society London B,* 274, 1237-1243.

Savarie, P.J., Shivik, J.A., White, G.C., Hurley, J.L. & Clark, L. (2001). Use of acetaminophen for large-scale control of brown treesnakes. *Journal of Wildlife Management*, 65, 356-365.

Schoener, T.W., Spilier, D.A. & Losoa, J.B. (2001). Predators increase the risk of catastrophic extinction of prey populations. *Nature*. 412, 183-186.

Schwartz, A. & Henderson, R.W. (1991). *Amphibians and reptiles of the West Indies: descriptions, distributions and natural history*. Gainesville, Florida, University of Florida Press.

Schwartz, A. & Thomas, R. (1975). A checklist of West Indian amphibians and reptiles. *Carnegie Museum of Natural History, special publication,* 1,127-129.

Sementelli, A., H.T. Smith, W.E. Meshaka, Jr. & R.M. Engeman. (2008). Just Green Iguanas? The Associated Costs and Policy Implications of Exotic Invasive Wildlife in South Florida. *Public Works Management and Policy*, 12, 599-606.

Seward, N., VerCauteren, K., Witmer G. & Engeman, R. (2004). Feral swine impacts on agriculture and the environment. *Sheep and Goat Research Journal,* 19, 34-40.

SFWMD (South Florida Water Management District) (2008). *Executive Summary 2008 South Florida Environmental Report*. West Palm Beach, Florida, South Florida Water Management District.

Shigesada, N. & Kawasaki, K. (1997). *Biological Invasions: Theory and Practice*. Oxford, UK, Oxford University Press.

Simberloff, D. (2003). How much information on population biology is needed to manage introduced species? *Conservation Biology* ,17, 83-92.

Smith, H.T. & Engeman, R.M. (2004). A review of the colonization dynamics of the northern curly-tailed lizard (Leiocephalus carinatus armouri) in Florida. *Florida Field Naturalist,* 32, 107-113.

Smith, H.T. & Engeman, R.M. (2007). Leiocephalus carinatus armouri (northern curlytail lizard). Attack of northern mockingbird; assertion displays. *SSAR Herpetological Review,* 38, 457.

Smith, H.T., Engeman, R.M. & Meshaka, Jr, W.E. (2006). Leiocephalus carinatus armouri (northern curlytail lizard) predation. *SSAR Herpetological Review*, 37, 224.

Smith, M.M., Smith H.T. & Engeman, R.M. (2004). Extensive contiguous north-south range expansion of the original population of an invasive lizard in Florida. *International Biodeterioration and Biodegradation,* 54, 261-264.

Snow, R.W. (2006). Disposable pets, unwanted giants: pythons in Everglades National Park. In *Florida Exotic Plant Council, 21st Annual Symposium*, Gainesville, Florida, Florida Exotic Plant Council.

Stein, B.A. & Flack, R.S., editors (1996). *America's least wanted: alien species invasions of U.S. ecosystems*. Arlington, Virginia, The Nature Conservancy.

Towne, C.W. & Wentworth, E.N. (1950). *Pigs from Cave to Cornbelt.* Norman, Oklahoma, University of Oklahoma Press.

U.S. Congress. (1993). *Harmful Non-indigenous Species in the United States.* Washington D.C., Office of Technology Assessment, OTA-F-565, Government Printing Office.

USDA. (2002). *Environmental assessment: Management of predation losses to state and federally endangered, threatened, and species of special concern; and feral hog management to protect other state and federally endangered, threatened, and species of special concern, and candidate species of fauna and flora in the state of Florida.* Gainesville, Florida, United States Department of Agriculture, Animal Plant and Health Inspection Service, Wildlife Services.

USFWS (U.S. Fish and Wildlife Service). 2003. *Captive propagation and reintroduction plan for the Key Largo woodrat (Neotoma floridana smalli).* Vero Beach, Florida, U.S. Fish and Wildlife Service, South Florida Ecological Services Office.

USFWS (U.S. Fish and Wildlife Service). (1999). *Multi-species recovery plan for the threatened and endangered species of South Florida, vol. 1.* Vero Beach, Florida, US Fish and Wildlife Service.

Western, D. (1974). The distribution density, and biomass density of lizards in a semi-arid environment of northern Kenya. *East African Wildlife Journal*, 12, 49-62.

Wilcove, D.S., Rothstein, D, Dubow, J., Phillips, A. & Losos, E. (1998). Assessing the relative importance of habitat destruction, alien species, pollution, over-exploitation, and disease. *BioScience,* 48, 607-616.

Winter, L. (2004). Trap-neuter-release programs: the reality and impacts. Journal of *American Veterinary Medical Association*, 225, 1369-1376.

Winter, L. & Wallace, G.E. (2006). *Impacts of feral and free-roaming cats on bird species of conservation concern.* The Plains, Virginia, American Bird Conservancy.

Wood, G.W. & Barrett, R.H. (1979). Status of wild pigs in the United States. *Wildlife Society Bulletin,* 7, 237-246.

Woods, M., McDonald, R.A. & Harris, S. (2003). Predation of wildlife by domestic cats *Felis catus* in Great Britain. *Mammal Review*, 33, 174-188.

LIST OF CONTRIBUTORS

L. Agudo
Shuqing An
Miguel Vazquez Archdale
J.R. Arévalo
Brian L. Becker
Stephan G. Bullard
Mary R. Carman
Lydie Chapuis-Lardy
Bernice Constantin
Nicolas Dassonville
Kirk W. Davies
Zifa Deng
Sylvie Domken
Richard Engeman
D. Evans
J.D. Fiore
T.S. Fredericksen
Scott Hardin
Dustin D. Johnson
Weiguo Li
David P. Lusch
Pierre Meerts
Engin Meriç
Walter E. Meshaka, Jr.
A. Naranjo
W.D. Pitman
Jiaguo Qi
M. Salas
Henry Smith
M. Thurman
Nathan M. Torbick
Sonia Vanderhoeven
Jianbo Wang
Zhongsheng Wang

Mehmet Baki Yokeş
Yingbiao Zhi
Changfang Zhou

INDEX

C

D

E

F

G

H

I

N

O

P

Q

R

S

T

U

V

W

X

Y

Z